Water and Development

– A Composite View –

Water and Development
– *A Composite View* –

Editor
Dr. Mohinder Kumar Slariya
Environmental Sociologist
Govt. Post Graduate College, Chamba
Himachal Pradesh, India

BIOTECH
2013
BIOTECH BOOKS®

ISBN 978-81-7622-269-3

Published by: **BIOTECH BOOKS®**
4762-63/23, Ansari Road, Darya Ganj,
New Delhi - 110 002
Phone: +91-011-23262132
E-mail: biotechbooks@yahoo.co.in

Printed at: **Chawla Offset Printers**
Delhi - 110 052

PRINTED IN INDIA

Acknowledgements

I would like to express my sincere thanks to University Grants Commission who provide financial assistance for national conference, which was popularly announced as *Chamba Water Meet-2011* and collaborating organisation *Mountain Forum Himalayas* and *Sewa Himalayas*, organisations committed to aware people on environment, development and water issues from the last more than three decades. I would like to thanks all the scholars who make their presence possible in the conference and contributed their papers and allow me to publish their work in form of this book.

I am thankful to Dr. M.L. Sharma, Principal, Government PG College, Chamba for his continuous support and encouragement which has made this event possible. I am also thankful to the scientific committee of the conference who finalised the whole event by giving their scientific inputs.

First of all, informally, I feel indebted and I express my sincere thanks to Dr. RK Kaistha, my graduation teacher, Prof. SK Sharma, Prof. OP Monga and Prof. Mohan Jharta my post-graduation teachers. It is the outcome of the research taught by Prof. SK Sharma at master level in 1995, after which I develop some understanding on research and contacting research even today. I am thankful to all my colleagues (teaching and non-teaching), volunteers Sh. Mrinal Sharma, Deepak Joshi, Vishal Ahuja, Dharam Chand for their contribution to make the conference happen. I am thankful my family members, my mother, wife, son and daughter for their patience and compromises they had made during the time of conference. I am also thankful national and international organisations who accepted my papers and encourage me took such initiatives in the far-flung area, lacking almost every modern facility.

The responsibility for any mistake and for the ideas expressed herein is mine. I would be very happy to hear from you at mkslariya@gmail.com for any comment.

Mohinder Kumar Slariya

Preface

With the declaration of 2005-2015 as decade of *Water for Life International Decade for Action*, with the primary objective to promote efforts to fulfil international commitments on water related issues, these issues got momentum and drawn international attention. These commitments include the Millennium Development Goals to reduce by half, the proportion of people without access to safe drinking water by 2015, and to stop unsustainable exploitation of water resources. Two other goals identified at the 2002 World Summit on Sustainable Development were also adopted as part of the *Decade for Action:* halve, by 2015, the proportion of people who do not have access to basic sanitation; and develop Integrated Water Resource Management and water efficiency plans. Moreover, concerns regarding water availability, clean water, sanitation and sustainable development have been voiced throughout the world, particularly against the background of vulnerabilities in context of climate change.

This book is a collection of 18 invited papers that address water and development issues in context of climate change with a special focus on South Asia region. Scholars from two countries India and Nepal have contributed in the UGC sponsored National Conference organised by the Department of Sociology, Govt. PG College, Chamba, popularly known as *Chamba Water Meet-2011.* The conference has been organised in collaboration with *Mountain Forum Himalayas*, an organisation is committed for water and environmental issues in Himalayan region. The content of the conference and book is close to the United Nation's *Decade for Action* and highlighted by the scholars, planners, executors. I do hope you find it both interesting and informative.

About the Contents

Bishnu Pandey and Suman Shakya's paper *Determination of Scaling Tendency and Corrosivity of Rural Drinking Water Using Langlier's Saturation Index* is the lead-off for the book. This paper provides an extensive overview of the rural drinking water in rural areas of Nepal. The second paper is detailed description of *Effective Water Resources Management & Agriculture Development in Upper Ganga - Yamuna plain* by Dr. AP Singh. It is case study of upper Ganga region which describes the use and management of water resources for agriculture purposes in wake of climate change and highlights the threats being faced by agriculture and also highlights measures to cope the situation. The next chapter has been contributed by executive engineer cum researcher Mr. Abdul Khan and he has made an attempt to highlight use of *water for agriculture for securing livelihood and food security*. The fourth chapter is case study of Langtang Basin in Nepal describing the *impacts of climate change on snow melt runoff* by Bikas Bharrarai and Regmi of central department of environmental science, Tribhuvan University, Nepal. By using snow runoff model (SRM) and Moderate Resolution Imaging Spectro radiometer (MODIS) contributors has analysed situation of snow melt in Nepal region of Hind-Kush Himalayas.

Next paper entitled, *Bringing relevance to agriculture in hills in the face of emerging natural and human challenges* has been contributed by Dr. Punam Sharma and Sanjeev Vasudeva. The authors have highlighted the challenges faced by the farmers and suggestions to cope with the situation and ensure the food security. Attempt has been made to provide an alternative in form of conservative agriculture also highlight efforts of state and central government in this regard. Case study Khajjiar Wetland conducted by Dr. Dinesh Kumar highlighting the *Impacts of Climatic Change on Himalayan Wetlands: Adaptation and Management* strategies to manage such wetlands in hills particularly and at global level in general. *Scramble for Water in Himalayas: A Case Study of Water Looting in Ravi basin in Himachal Himalayas* is the next paper contributed by Dr. Mohinder

Slariya. This chapter highlights mad scramble of water looting in peer panjal range of western Himalayas based on the micro level study conducted in saal valley in chamba district of Himachal Pradesh and also highlight multifarious impacts of such scrambles on the lives and livelihood of the people at micro level. Contribution of Dr. Krupesh Chauhan and his students in form of chapter entitled, *Remedial Measure of Tapi River Flood at Surat city,* is an attempt at micro level to understand the problems related with flood in the western part of India in the state of Gujarat. Contributors tried to identify threats and scenarios and try to present scenario before readers the measures to articulate the disaster recovery strategies with concenterating on four elements of the generic disaster recovery plan: Mitigation, preparedness, response, and recovery with respect to floods. The paper also try to provide examples of software that can help disaster recovery professionals in the planning and implementation process. *Impacts of Land Use and Climate Change Scenarios on the Water Flux: A Case Study of Sal River Watershed, Chamba (H. P.)* is the next paper contributed by Dr. Kuldeep Pareta et.al. The paper is a detailed description of impact of climate change on the land use pattern and water flux in saal valley in chamba district of Himachal Pradesh, India by using semi-distributed hydrological model Soil and Water Assessment Tool (SWAT), IRS-P6-LISS-IV-Mx (5.8m) and IRS-P6-LISS-III (23.5m) satellite imageries. The chapter is aiming to map the existing field scale land use practices in order to determine their impact on rainfall and temperature variations based on the Intergovernmental Panel on Climate Change (IPCC) projections at micro level. Next chapter is an attempt of retired engineer Sh. Ramesh Ganeriwal from USA highlighting farmers' perspective at micro level and the name of paper is *Direct Negative Impacts of Micro-Hydel Projects on Poor Farmers.*

Design Principles for Community –Based Natural Resource Management: A Case of Community Managed Canal Irrigation System by Sachin Kumar and Shailendra Singh Bisht is the next micro level study conducted at Dehradoon, Uttaranchal, India. Concomitant changes in water quantity and quality due to climate change are responsible to increase the vulnerability of poorer sections of society in general and farmers in particular. Paper is boost to devise adaptive strategies to mitigate the vulnerabilities by ensuring sustainable water resource management. Sh. Vivek Trivedi continue the theme and content of the book with his contribution, *Climate Change and*

Water Resources: Issues and Options. He emphasise in this chapter on unplanned development, concrete surfacing, reclaiming of land and other human induced action that has impinged water bodies, river course, catchments. Mainly human factors are responsible for breaking the silence of nature; troubling flora and fauna, littering non-biodegradable waste in the pristine glaciers, throwing untreated sewage, industrial waste chemical in the sacred perennial river originating from great Himalayas. He also stressed on the skimpy implementation of rehabilitation mechanism due to bureaucratic inertia and lack of political commitment are hampering the success of pilot project initiated and in replicating the successful models for the conservation of natural resources viz. frozen glaciers, dense forests, wild life (flora-fauna) etc. *Sustainable Development in Context of climate change* is the contribution of Dr. PK Gupta. He highlighted the concept of sustainability and co-relates it with the modern theories of development and propounded the concept aqua care with an advocacy that it should be included in the policies and curriculum in form of EDUCARE. *Community Management of Groundwater- Basis for Piots,* is the next chapter contributed by Sh. Amit Upmanyun, social scientist from *Advanced Center for Water Resources Development And Management, Pune, India.* Ground water is considered common pool resource like other sources like water, forests, rivers, common land etc. broadly speaking having common ownership or use. As a common pool resource, chapter explains the need of community management of this resource and also suggest some protocols and management strategy for community based groundwater management.

There is a shift from water issues to issue of industrialization and presented in form of next chapter entitled, *Industrialization and Environment- A Sociological Analysis.* The chapter has been contributed by Prof. Ashwani Kumar and Dr. Mohinder Slariya with a focus on ACC cement factory at Barmana in Bilaspur district of Himachal Pradesh, India. The chapter highlight the developmental processes in Shivalik Himalayas with a special focus on disturbances in ecological and environmental aspects and vital disturbance it has created in the physio-socio-cultural and non-material environment. The paper is based on a study conducted in five villages situated in semi-circle situation and also surrounded by Kol Dam, Dehar dam and Govind Sagar Lake.

Climate Change and its Impact on Seasonal Streams Flowing in the Natural Forest Areas of Central India, finest research based contribution of Dr. Manish Mishra, is an attempt of the contributor to see the impact of climate change on the seasonal streams flowing in the natural forests of one of richest natural resources state i.e. Madhya Pradesh in central India. This paper raised the issue of availability of water to the tribal women, who are forced to fetched water from minimum 3-5 kms daily because of the impacts of climate change. An attempt has also been made to highlight the importance of conserving natural flowing streams located in the forests to provide clean and abundant water for use for the tribal.

Background and Readership: why this book has been written?

Relationship with Other Books: try to point out earlier work and how work is different that of others...

Mohinder Kumar Slariya

Contents

Chapter 1

Determination of Scaling Tendency and Corrosivity of Rural Drinking Water Using Langlier's Saturation Index

☆ *Bishnu Pandey and Suman Shakya*

ABSTRACT

This chapter is based on the study carried out with an objective to assess the scaling tendency and corrosivity of rural drinking water sources (springs) of the 9 hilly districts of Central Development Region, Nepal. A total of 113 water samples collected during August 2009 were subjected to water quality analysis for the determination of Temperature, pH, Total Dissolved Solids (TDS), Total Alkalinity (TA) and Calcium Hardness (CaH). Scaling tendency and corrosively of these samples were then determined using Langlier's Saturation Index (LSI). Majority of the samples were found to be corrosive. Only 27 per cent (30 out of 113) of the spring samples were found to be scaling forming. Two of them were found to be neutral and 72 per cent (81 out of 113) of the samples were found to be corrosive. None of the spring samples violated

WHO GV and Nepali standards for Temperature and TDS whereas TA, CaH and pH violation occurred in 11 per cent, 0.9 per cent, and 5.3 per cent of the samples respectively.

Keywords: *Scaling tendency, Corrossivity, Rural drinking water, Langlier's Saturation Index, WHO.*

Introduction

Majority of the rural population in the hilly region of Nepal depend heavily on gravity fed water supply system for drinking purpose. Often times, the real problem with the existing water supply system in this region is not a lack of safe drinking water but calcified pipes that clog or retard the flow of water causing failure of the water supply system in the long run. Besides, the scales, typically a carbonate residue, cause build up on hot water heaters, impart an alkali taste to the water, reduce the efficiency of the water heaters, and cause other aesthetic problems. On the other hand, Corrosive water can react with the household plumbing and metal fixtures resulting in the deterioration of the pipes and increased metal content of the water. This metal loading can significantly decrease the quality of the drinking water. Lead and cadmium, two potentially toxic metals, are attributable almost entirely to leaching caused by corrosion of distribution pipes (Schock 1999). Studies have shown that corrosion products attached to pipe surfaces or accumulated as sediments in distribution system can shield microorganisms from disinfectants (Geldreich and Lechevallier 1999). These microorganisms can cause many problems such as slimes, bad odour and taste (Bina and Porzamani 2005).

Various models and indices such as Langelier Saturation Index (LSI) and the Ryznar Stability Index (RSI) have been suggested to predict the scale forming potential of an aqueous solution based on some easily measurable hydraulic and pipe physical parameters. LSI is a qualitative index to determine if the solution is under saturated or supersaturated with $CaCO_3$ that is directly related to the corrosive rate of metal pipe exposed to water (Siobhan *et al.*, 2002). LSI is a system for estimating or predicting the amount or degree of problems with lime scale, a particular water supply will cause. The LSI value usually lies between –3 and +3. If the LSI is positive, then the water will deposit $CaCO_3$; on the other hand, the negative LSI value makes the $CaCO_3$ to dissolve in water (Langelier 1946).

The present study is an attempt to assess the scaling potential and corrossivity of the water of the proposed drinking water supply sources using Langelier Saturation Index (LSI).

Materials and Methods

A total of 113 samples collected from springs during August 2009 most of which represent hilly districts (Ramechhap, Sindhuli, Makwanpur, Lalitpur, Kavre, Sindhupalchowk, Nuwakot, Dhading and Chitwan) were subjected to water quality analysis. All the physico-chemical analyses were carried out insitu.

Table 1.1: Materials and Methods

Sl.No.	*Parameters*	*Units*	*Methods*	*Instruments*
1.	Temperature	°C	Thermometric	Thermometer
2.	pH	-	Instrumental, as per Manufacture's manual	Portable HANA pH meter
3.	TDS	mg/L	Instrumental, as per Manufacture's manual	Portable HANA Conductivity meter
4.	Total Alkalinity	mg/L	Titrimetric, as per the Standard method	ENPHO Kits
5.	Ca Hardness	mg/L	Titrimetric, as per the Standard method	ENPHO Kits

Scales forming tendency of the water samples were determined using Langelier's Saturation Index (LSI) following prescribed standard methods (Langlier, W.F., 1946)

$$LSI = pH - pH_s$$

where,

$$pH = -\log [H^+] \text{ and } pH_s = (9.3 + a + b) - (c + d)$$

where,

$a = (\text{Log}_{10} [TDS] - 1)/10$

$b = -13.12 \times \text{Log}_{10} (°C + 273) + 34.55$

$c = \text{Log}_{10} [Ca^{2+} \text{ as } CaCO_3, mg/L] - 0.4$

$d = \text{Log}_{10} [\text{alkalinity as } CaCO_3, mg/L]$

If LSI > 0, scale forming occurs, if LSI < 0, water is corrosive and if LSI = 0, water is neutral to Scale forming (Langlier, W.F., 1946 and Rafferty, K., 2003).

All the statistical analyses were carried out using MS Excel-2007.

Result and Discussion

None of the samples violated WHO GV and Nepali standards for TDS. 5.3 per cent of the samples violated the standards for pH. All of those that violated the standards for pH had pH > 8.5. Similarly, 11 per cent of the samples violated the standards for Total Alkalinity (TA) and only 0.9 per cent of them violated the standards for Calcium Hardness (CaH).

The mean Temperature was found to be 20.40 ± 0.80 °C. Minimum and Maximum Temperature were found to be 12.0 °C and 30.0 °C respectively. The Nepali national water quality standards require pH to be between 6.5 and 8.5 (Government of Nepal, 2006). The mean pH was found to be 7.85 ± 0.09. Minimum and Maximum pH value were found to be 6.50 and 9.0 respectively. Damsidol WSS in Lalitpur district was the only scheme in the study area where pH value exceeded the upper limit of the WHO standard. The mean TDS value was found to be 82.0 ± 13.0 mg/L. Minimum and Maximum TDS values were 4.0 mg/L and 282.0 mg/L respectively. As per the TDS classification (Fetter 1990), 100 per cent of the water samples of the study area registered to fresh type (TDS < 1,000 mg/L). Fairly low TDS value as compared to WHO standard (1000 mg/L) for both types of water samples may be attributed to their low concentration of the dissolved salts. The Total Alkalinity of the water is its capacity to neutralize acids. The mean Total Alkalinity of gravity water samples was found to be 99.0 ± 15.0. Minimum and Maximum Total Alkalinity value were found to be 8.00 mg/L and 404.0 mg/L respectively. However 10.6 per cent (12 out of 113) of the samples has been found to exceed the limit (Table 1.4). Sindhuli, Chitwan and Makwanpur were the districts with relatively higher concentration of Total Alkalinity. This may be due to the higher inputs of carbonates and bicarbonates. Similarly mean Calcium Hardness was found to be 64.0 ± 11.0mg/L. Minimum and Maximum Calcium Hardness concentration were found to be 8.0 mg/L and 256.0 mg/L respectively.

Table 1.2: Statistical Summary of the Drinking Water Quality Parameters

Source Type	*Sample Size*	*Statistic*	*pH*	*Temp*	*TDS*	*TA*	*CaH*
Spring	113	Min	6.50	12.0	4	8	8
Water		Max	9.00	30.0	282	404	256
		Mean± CI	7.85± 0.09	20.4± 0.8	82± 13	99± 15	64± 11
		Median	7.90	21.0	48	72	40
		SD	0.48	4.3	72	83	58
		WHO GV*	6.5-8.5[a]	NA	1000	200	100-300
		NDWS*	6.5-8.5[a]	NA	1000	NA	200
		Per cent	5.3	0	0	11	0.9

TDS: Total Dissolved Solids; Temp: Temperature; TA: Total Alkalinity; CaH: Calcium Hardness.

All the parameters are in mg/L except pH, and Temp (in ºC).

'SD' indicates Standard Deviation, 'CI' indicates Confidence Interval.

*WHO–World Health Organization

*NDWS-Nepal Drinking Water Standards, 2062 and National Drinking Water Quality Standard Implementation Guideline, 2062 year: 2063(B.S.) Government of Nepal, Ministry of Physical Planning and Works, Singhadurbar, Kathmandu, Nepal.

[a] levels are minimum to the maximum

'per cent ' indicates the percentage of samples that exceeds the WHO GV for the given parameters.

Langelier's Saturation Index

Majority of the samples were found to be corrosive (LSI< 0) rather than scale forming (LSI >0) (Figure 1.6). 27 per cent (30 out of 113) of the water samples were found to be scale forming while 72 per cent (81 out of 113) of the samples were corrosive and 2 samples were neutral (LSI=0).

Higher scale forming tendency of the spring water sources were observed in Ghari Thunsi and Nimane Water Supply Schemes in Sindhuli district(LSI=1.1 and 1.2 respectively), Chainpur Scheme in Makwanpur district (LSI=1.1)and Kalika, Samfrang, and Bhalumare Jhul water supply Schemes in Chitwan district(LSI=1.0,1.2, and 1.0

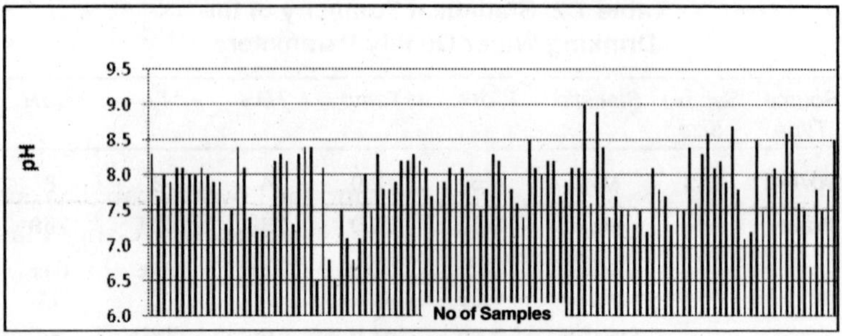

Figure 1.1: pH Variation of Spring Water Samples

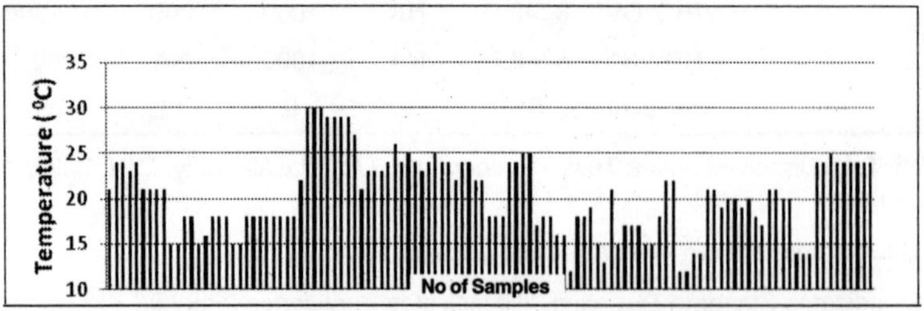

Figure 1.2: Temperature Variation of Spring Water Samples

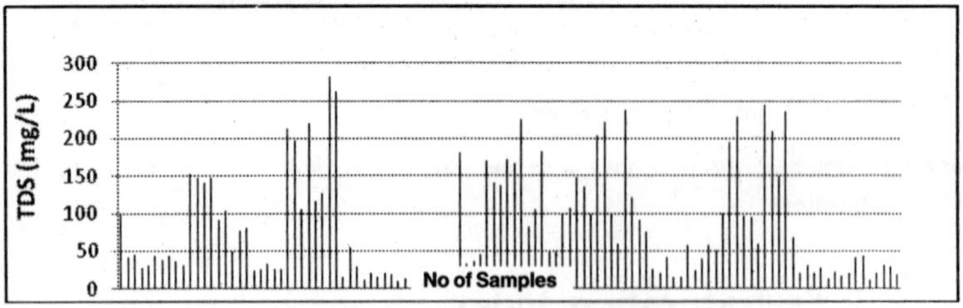

Figure 1.3: TDS Variation of Spring Water Samples

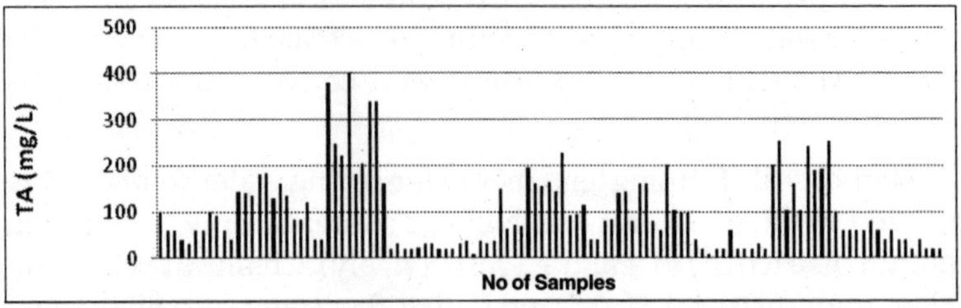

Figure 1.4: TA Variation of Spring Water Samples

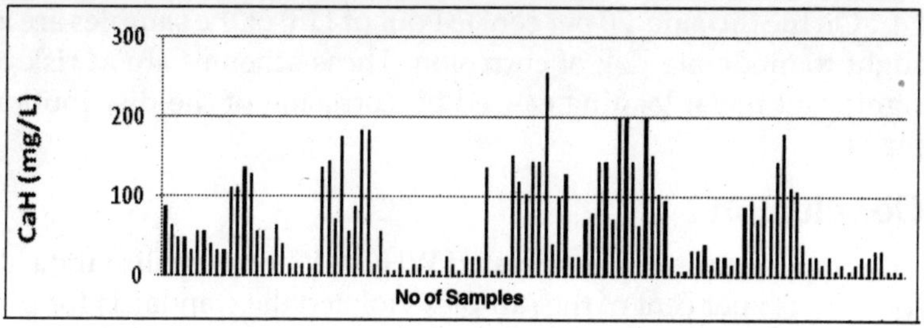

Figure 1.5: CaH Variation of Spring Water Samples

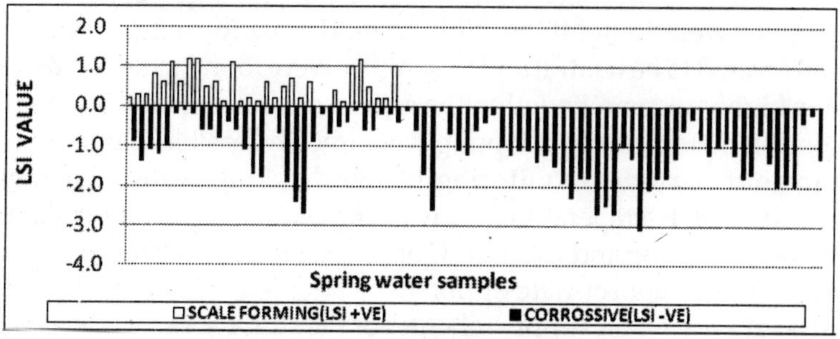

Figure 1.6: LSI of Gravity Water Samples

respectively). These schemes are at slight risk of $CaCO_3$ deposition in their pipelines.

Table 1.3: Remarks/Recommendation Based on LSI Categories

LSI Value (Exclusive)	*No. of Samples*	*Remarks/Recommendation*
0	2	Balanced, treatment not needed.
0 to 0.5	13	Slight scale forming, treatment not needed
0.5 to 1.0	10	Some faint coating treatment generally not needed
1.0 to 1.5	7	Mild scale forming, probable minor aesthetic problems
0 to -1.0	36	Mild corrosion, treatment probably not needed
-1.0 to -2.0	34	Mild corrosion, treatment should be considered
-2.0 to -3.0	10	Moderate corrosion treatment should be considered
-3.0 to -4.0	1	Moderate corrosion, treatment recommended

On the flip side, 70 per cent (81 out of 113) of the samples are at slight to moderate risk of corrosion. These schemes are at risk of significant metal loading caused by corrosion of the distribution pipes.

Conclusion

None of the samples violated WHO GV and Nepali standards for TDS. 5.3 per cent of the samples violated the standards for pH. Similarly, 10.6 per cent of the samples violated the standards for Total Alkalinity (TA) and 0.9 per cent of them violated the standards for Calcium Hardness (CaH). Out of 113 springs surveyed, 30 samples were found to be slight to mild scale forming, 2 samples were found to be neutral and 81 samples were found to be corrosive. It is highly recommended to the authorities to control scaling and corrosion problems by adopting economically feasible methods and carryout periodic monitoring of the drinking water schemes. Avoiding high temperature(by providing enough depth for pipelines) to control scaling and use of PVC or CPVC pipe to prevent corrosion might be the appropriate option to be adopted for these proposed rural gravity water supply schemes of the study area.

References

Bina, B. and Porzamani, H., 2005. Potential of corrosion in Baghdaran water resources, Esfahan, Mazandaran. *J. Environ. Health*, 3: 6–10.

Fetter, C.W., 1990. *Applied Hydrogeology*. CBS Publishers and Distributors, New Delhi, India.

Geldreich, E.E. and Lechevellier, M., 1999. Microbiological quality control in distribution systems. In: *Water Quality and Treatment: A Handbook of Community Water Supplies*, (Ed.) A.W.W. Association. McGraw-Hill, pp. 18–49.

Government of Nepal, National Drinking Water Standards, 2062 and National Drinking Water Quality Standard Implementation Guideline, 2062 year: 2063(B.S.) Government of Nepal, Ministry of Physical Planning and Works, Singhadurbar, Kathmandu, Nepal

Langelier, W. F., 1946. Chemical equilibria in water treatment. *Journal of American Water Works Association*, 38: 169.

Rafferty, K., 2003. Scaling in geothermal heat pump systems. U.S. Department of Energy (July, 1999) Available: http://geoheat.oit.edu/otl/scaleghp.pdf.

Schock, M.R., 1999. Internal corrosion and deposition control. In: *Water Quality and Treatment: A Handbook of Community Water Supplies,* (Ed.) A.W.W. Association. McGraw-Hill, pp. 1–17.

Chapter 2

Effective Water Resources Management and Agriculture Development in Upper Ganga: Yamuna Plain

☆ *Anirudh Prasad Singh*

Water resources are sources of water that are useful or potentially useful to humans. Uses of water include agricultural, industrial, household, recreational and environmental activities. Virtually all of these human uses require fresh water.

The 70 per cent of earth's surface is covered by oceans, as 3 per cent of fresh water and ¾ as frozen. The world is facing water shortage and other water resource problems. There is need of water resource management to reduce the water resource problems.

LULUCF has impacts on the global carbon cycle and as such these activities can add or remove carbon dioxide (or, more generally, carbon) from the atmosphere, contributing to climate change. LULUCF has been the subject of two major reports by the Intergovernmental Panel on Climate Change (IPCC). Additionally, land use is of critical importance for biodiversity.

Agriculture development means providing assistance to the crop producers with the help of various agricultural resources.

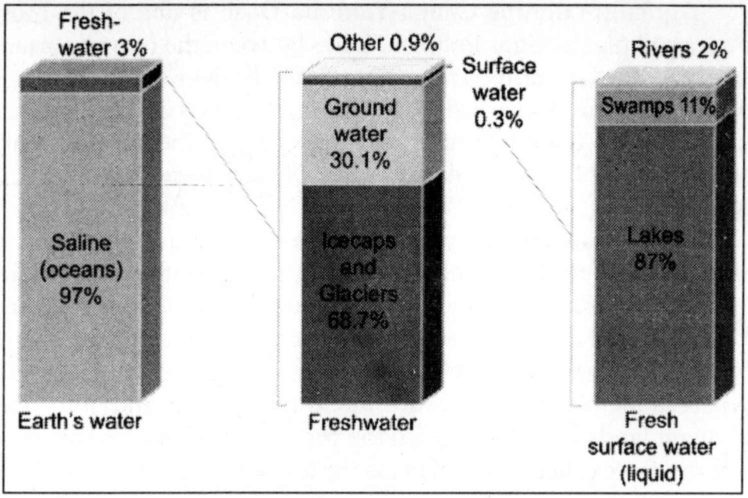

Figure 2.1: Distribution of Earth's Water

Providing protection, assisting in the research sphere, employing latest techniques, controlling pests and facilitating diversity all fall within the purview of agriculture development.

The study area flows through the state of Uttar Pradesh and is known as the Upper Ganga – Yamuna Doab. The region is essentially a most fertile tract lying between the two master streams of western Uttar Pradesh, the Ganga and the Yamuna making the eastern and western boundaries of the study area. The region embraces Haridwar, Saharanpur, Muzaffarnagar, Meerut, Baghpat, Ghaziabad, Gautambudha Nagar and Bulanshahr districts of Uttara Khand and western Uttar Pradesh, comprising an area of 20738·97 km^2 with general density of 935·17 persons/km^2. The study area extends between 28° 8/ to 30° 27/22// north of latitudes and 77° 00/35// to 78° 45/ east of longitudes. The region is well covered by good network of national and state highways and rail lines.

As a matter of fact, the region is a part of the vast tract of Indo-Gangetic Plain built up of alluvial sediments. The entire region is a monotonous level plain with slight undulations towards the inner part with little relief. It is a gradational plain composed of the sediments deposited by Ganga-Yamuna Rivers and, thus, is literally, the "dust of the mountains."

Agriculturally the Ganga-Yamuna Doab is one of the most important part of Uttar Pradesh. It lies between the two important streams – the Ganga and the Yamuna, which cover an area of about 58,400 km^2. The Doab consists of 23 districts which can be classified into three sub-regions namely, the Upper Doab, the Central Doab, and the Lower Doab. The entire region of the Ganga-Yamuna Doab forms a part of the Ganga Plain. The alluvial deposits brought down by the Himalayan Rivers spread over the entire area and the general slope of the plain very gently runs towards the eastern parts of the state of Uttar Pradesh. The alluvial deposits of the Ganga-Yamuna Doab can be classified into two categories: (i) the khadar lands, and (ii) the bhangar lands. The khadar lands contain newer alluviums, whereas the bhangar lands are made up of older alluvium. The bhangar lands are marked with the presence of saline and alkaline efflorescence called 'reh', whereas the khadar lands are free from it.

There are 13 major crops grown in the Ganga-Yamuna Doab and all of these together accounted for 94.92 per cent of the total cropped area. For the purpose of analysis these crops were grouped into four broad categories: (a) cereal crops to include rice wheat, maize, barley, jawar and bajra, (b) pulses to include gram, black gram (urad) green gram (moong) lentil (masoor) peas and pigeon pea, (c) oilseeds to include mustard, linseed seasum (til), groundnut and sunflower (d) cash crops to include sugarcane and potatoes.

Agriculture occupies an important place in the economy of Uttar Pradesh and more especially in the Ganga-Yamuna Doab, it not only provides food to the people but also raw materials to numerous agro-based industries. About 70 per cent of the working population is directly or indirectly engaged in agriculture.

In last three decades agriculture has shifted from subsistence to commercial agriculture. The Ganga-Yamuna Doab is one of the fertile regions of the state of Uttar Pradesh, but comparatively the yields of crops are below than the states of Punjab and Haryana of the country.

So there is a considerable scope for the improvement of agricultural production and productivity per hectare and per agricultural worker, particularly on small and marginal farms which will help to increase income levels and improve the quality of life of the people in the rural areas.

The Concept of Water Stress is Relatively Simple

According to the World Business Council for Sustainable Development, it applies to situations where there is not enough water for all uses, whether agricultural, industrial or domestic. Defining thresholds for stress in terms of available water per capita is more complex, however, entailing assumptions about water use and its efficiency. Nevertheless, it has been proposed that when annual per capita renewable freshwater availability is less than 1,700 cubic meters, countries begin to experience periodic or regular water stress. Below 1,000 cubic meters, water scarcity begins to hamper economic development and human health and well-being.

The trend towards urbanization is accelerating. Small private wells and septic tanks that work well in low-density communities are not feasible within high-density urban areas. Urbanization requires significant investment in water infrastructure in order to deliver water to individuals and to process the concentrations of wastewater – both from individuals and from business. These polluted and contaminated waters must be treated or they pose unacceptable public health risks.

Cropping activities in Upper Ganga–Yamuna plain go on all the year-round as water is available for crops. In northern India, there are two distinct seasons, kharif (July to October), and Rabi (October to March). Crops grown between March and June are known as Zaid. In some parts of the country, there are no such distinct seasons, but there they have their own classification of seasons. The village revenue officials keep plot-wise record of crops grown in each season. These are annually compiled district-wise, state-wise and on all-India basis. From these records one could calculate the relative abundance of a crop or a group of crops in a region. The Government of India places high priority on reducing poverty by raising agricultural productivity. However, bold action from policymakers will be required to shift away from the existing subsidy-based regime that is no longer sustainable, to build a solid foundation for a highly productive, internationally competitive, and diversified agricultural sector.

Agricultural productivity is measured as the ratio of agricultural outputs to agricultural inputs. While individual products are usually measured by weight, their varying densities make measuring overall agricultural output difficult. Therefore, output is usually measured as the market value of final output, which excludes intermediate products such as corn feed used in the meat industry. This output value may be compared to many different types of inputs such as labour and land (yield). These are called partial measures of productivity.

The main approaches of water resources management are:

- ☆ Increase the usable supply.
- ☆ Decrease the unnecessary loss of water.

For this the following steps are to be taken:

Dams and Reservoir

Rain water and water from show can be captures by dams and reservoirs for use during dry seasons. Dams also control the flow and reduce the danger of flooding, provide irrigation and generate relatively cheap electricity.

Water Diversion

In many developed countries fresh water has been diverted from water rich areas with scarcity of water in order to meet the water crisis.

Tapping Ground Water

When surface resupplies are insufficient the water can be trapped from the ground water. Any underground geological formation that contains enough ground water to supply human needs is called an aquifer. However, the increased use of ground water gives rise to several problems like:

- ☆ Depletion when ground water is withdrawn faster than it is recharged by precipitation.
- ☆ Subsidence or sinking of the ground as the ground water is withdrawn.
- ☆ Salt water is withdrawn faster than it is recharged.
- ☆ Ground water is contaminated by human activities

Reuse of Water

- ☆ **Inadvertent Reuse:** The water is withdrawn from the river used treated and discharged to river.

- **Indirect Water Reuse:** It is a planned endeavour through waste water reclamation cycle.
- **Direct Water Reuse:** It refers to treated water that is piped directly to the next user, *i.e.* for industrial or agriculture activities

A number of schemes for intensive cultivation of rice, wheat, sugarcane, maize, jawar, pulses, and oilseeds have been put into operation.

Table 2.1: Methodology for the Calculation of Crop Yield Index

Name of Crop	*Yield (qnt/ha)*		*Area of Crop in the District (000 ha)*	*Crop Yield in the District as Percentage of the State*	*Percentage Multiplied by Area (m ha.)*
	Average Yield in the State	*Yield in the District*		*Col. 3/Col. 2 × 100*	*Col. 4 × Col. 5*
1	*2*	*3*	*4*	*5*	*6*
Rice	18.72	17.83	129.50	94.94	11820.03
Wheat	25.25	27.47	2304.01	108.79	250653.25
Maize	16.30	17.64	440.28	108.32	47647.10
Jowar	11.26	7.25	6.29	64.39	405.01
Bajra	11.94	12.62	898.49	105.70	94970.39
Barley	23.60	27.23	377.93	115.38	43605.56
			4151.50	449101.34	

Productivity Regions – Based on Cereal Crops

Cereals are the most important crops grown in the Ganga-Yamuna Doab to cover an area of 4,705 thousand hectares (71.26 per cent) of the total cropped area. Rice and wheat together cover 51.66 per cent of total cultivated area.

In fact there are seven districts which form high productivity region. These districts are namely, Saharanpur, Muzaffarnagar, Meerut, Bulandshahar, Ghaziabad, Mathura and Agra. Here the high productivity indices are due to several institutional and technological factors which have contributed a lot in last three decades. The districts characterized with medium productivity category are namely, Firozabad, Mainpuri, Etawah and Kanpur (rural).

Economic Considerations

Water supply and sanitation require a huge amount of capital investment in infrastructure such as pipe networks, pumping stations and water treatment works. It is estimated that Organisation for Economic Co-operation and Development (OECD) nations need to invest at least USD 200 billion per year to replace aging water infrastructure to guarantee supply, reduce leakage rates and protect water quality.

International attention has focused upon the needs of the developing countries. To meet the Millennium Development Goals targets of halving the proportion of the population lacking access to safe drinking water and basic sanitation by 2015, current annual investment on the order of USD 10 to USD 15 billion would need to be roughly doubled. This does not include investments required for the maintenance of existing infrastructure.

Once infrastructure is in place, operating water supply and sanitation systems entails significant on-going costs to cover personnel, energy, chemicals, maintenance and other expenses. The sources of money to meet these capital and operational costs are essentially either user fees, public funds or some combination of the two.

But this is where the economics of water management start to become extremely complex as they intersect with social and broader economic policy. Such policy questions are beyond the scope of this article, which has concentrated on basic information about water availability and water use. They are, nevertheless, highly relevant to understanding how critical water issues will affect business and industry in terms of both risks and opportunities.

Key Challenges

Water resources management is a highly political topic and planning and implementation needs to explicitly address the political economy of reform. This means recognizing that solutions have to be tailored to specific, widely-varying, circumstances. The art of reform is in picking the low-hanging fruit first, not making the best the enemy of the good, recognizing broader reforms outside of the water sector (often relating to overall economic liberalization, fiscal and political reform), and recognizing that those who are willing to change must be supported.

The financing needs for water-related infrastructure in developing countries are huge. The World Commission on Water has estimated that investments in water infrastructure in developing countries need to increase from the current level of about $75 billion to $180 billion a year over the next 25 years.

Land Use Change

Land use, land-use change and forestry (LULUCF) is defined by the UN Climate Change Secretariat as "A greenhouse gas inventory sector that covers emissions and removals of greenhouse gases resulting from direct human-induced land use, land-use change and forestry activities.

LULUCF has impacts on the global carbon cycle and as such these activities can add or remove carbon dioxide (or, more generally, carbon) from the atmosphere, contributing to climate change. LULUCF has been the subject of two major reports by the Intergovernmental Panel on Climate Change (IPCC). Additionally, land use is of critical importance for biodiversity.

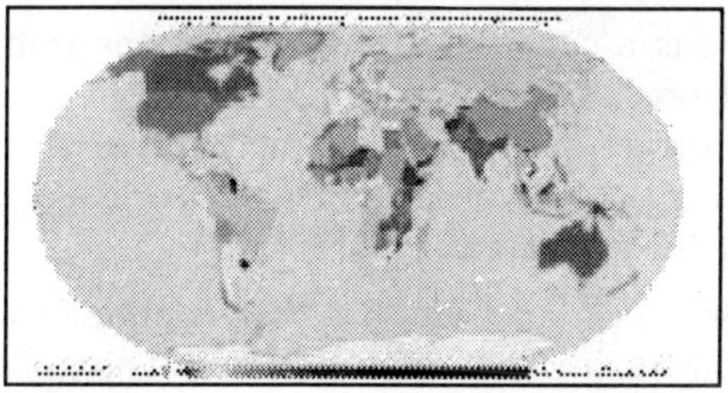

Figure 2.2: Climatic Impacts of Land-Use, Land-Use Change and Forestry

Per capita greenhouse gas emissions by country including land-use change Land-use change can be a factor in CO_2 atmospheric concentration, and is thus a contributor to climate change. IPCC estimates that land-use change (*e.g.* conversion of forest into agricultural land) contributes a net 1.6 ± 0.8 Gt carbon per year to the atmosphere. For comparison, the major source of CO_2, namely emissions from fossil fuel combustion and cement production amount to 6.3 ± 0.6 Gt carbon per year. This decision sets out the rules that govern how Kyoto Parties with emission reduction commitments (co-called Annex 1 Parties) account for changes in carbon stocks in land use, land-use change and forestry. It is mandatory for Annex 1 Parties to account for changes in carbons stocks resulting from afforestation, reforestation and afforestation and voluntary to account for emissions from forest management, cropland management, grazing land management and re-vegetation.

The rules governing the treatment of land use, land-use change and forestry for the second commitment period are currently being renegotiated as part of the Bali Action Plan under the Ad-Hoc Working Group on Further Commitments for Annex 1 Parties under the Kyoto Protocol (AWG-KP)

Agriculture occupies an important place in the economy of Uttar Pradesh and more especially in the Ganga-Yamuna Doab, it not only provides food to the people but also raw materials to numerous agro-based industries. About 70 per cent of the working population is directly or indirectly engaged in agriculture.

In last three decades agriculture has shifted from subsistence to commercial agriculture. The Ganga-Yamuna Doab is one of the fertile regions of the state of Uttar Pradesh, but comparatively the yields of crops are below than the states of Punjab and Haryana of the country.

So there is a considerable scope for the improvement of agricultural production and productivity per hectare and per agricultural worker, particularly on small and marginal farms which will help to increase income levels and improve the quality of life of the people in the rural areas.

Agricultural productivity has been computed worldwide by many scholars using different methods. Thompson measured relative productivity of British and Danish farms and express it in terms of gross output of crops and livestock. Kendall treated agricultural productivity as a mathematical problem and initiated a system of four coefficients (i) productivity coefficient, (ii) ranking coefficient, (iii) money value coefficient, and (iv) energy coefficient.

Stamp applied Kendal's ranking coefficient technique on an international level to determine agricultural efficiency in a number of countries and by selecting some major crops. Stamp suggested a method for measuring the agricultural productivity by converting the total agricultural productivity in calories. Shafi applied the technique of 'ranking coefficient' of Kendall for measuring the agricultural. The average amount of annual rainfall received in India is about 105 cm, which is equivalent to about 3000 trillion cubic meters of water. About half of this amount is available as surface water in which Ganga River carries about 460 billion m^3. The replenish able water component of groundwater potential of the country has been assessed at 456,000 mcm^2 out of this 9287 mcm^2 lies in the upper doab of Ganga – Yamuna even then the tight cornerness due to mismanagement of irrigation water is a natural challenge for the farmers, therefore, the battle against environmental degradation and depletion of water resources is a part of unceasing struggle for agricultural development.

Water is a delectable resource of a very sensitive nature. It is over exploitation may prove counter productive in the long run. Because of increasing demand, the over pumping of groundwater aquifers and dangerous exploitation of surface water may result a drastic fall in water resources and consequences thereby, will affect the agricultural production. Due to environmental changes and syncope in water resources, a problem of land use and agricultural development inverse the consequences of the regional phenomenon. Adequate and assured availability of irrigation is a crucial determinant of agricultural productivity, therefore, providing security to agriculture, groundwater has distinct advantages over surface water and in years to come it is destined to become the mainstay of agricultural economy. The natural phenomenon which occurs in proximity and pose of the area of study is a threat to land use and agriculture, therefore, the present study is an emphasis of technological modification in the use of land and agricultural operations to redress the regional inequalities in the level of agricultural development and management of water resources.

The study area flows through the state of Uttar Pradesh and is known as the upper Ganga – Yamuna Doab. The region is essentially a most fertile tract lying between the two master streams of western Uttar Pradesh, the Ganga and the Yamuna making the eastern and western boundaries of the study area. The region embraces Haridwar, Saharanpur, Muzaffarnagar, Meerut, Baghpat, Ghaziabad, Gautambudha Nagar and Bulanshahr districts of Uttara Khand and western Uttar Pradesh, comprising an area of 20738·97 km^2 with general density of 935·17 persons/km^2. The study area extends between 28° 8′ to 30° 27′22″ north of latitudes and 77° 00′35″ to 78° 45′ east of longitudes. The region is well covered by good network of national and state highways and rail lines.

As a matter of fact, the region is a part of the vast tract of Indo-Gangetic Plain built up of alluvial sediments. The entire region is a monotonous level plain with slight undulations towards the inner part with little relief. It is a gradational plain composed of the sediments deposited by Ganga-Yamuna Rivers and, thus, is literally, the "dust of the mountains."

Economic and Policy Issues

Government policies and regulations on water management have so far been unable to stem the growing problems related to water quality and quantity in India. For the most part, this is due to the lack of implementation and enforcement of the existing regulations. The National Water Policy (NWP), which was adopted in 1987, is the principal document stating the government's position on water development issues. The government's commitment to holistic development of the country's water resources is well stated, but so far little has been done to actually employ these recommendations in the decision-making process. Similarly, enforcement of the regulations governing the development and protection of water resources has been poor and serious abuses continue to occur throughout the country. More stringent and enforceable regulations need to be put into place to prevent further degradation and wastage.

Economic incentives and subsidies have not been beneficial in terms of promoting conservation of water resources. Government subsidies to the irrigation and domestic water supply sectors in the past several decades have allowed for greater expansion and development of both, but have also led to a complete under-valuing of the resource. For surface and groundwater irrigation, charges are not based on the unit of volume consumed and in many cases, have not been revised for several years. Groundwater extraction by pumping is indirectly encouraged through subsidies for fuel and electricity (World Bank 1999). Rural and urban water charges are much lower than the cost of provision and suffer from poor operation and maintenance. The result in all sectors has been higher consumption and insufficient cost recovery.

Effective economic and management policies are needed to prevent the crisis that threatens India in the coming years. Good management of the country's water resources will effectively reduce the amount of pollution and over-exploitation that is currently plaguing the nation's surface, ground and coastal waters. The consequent improvement in water quantity and quality will also have repercussions in terms of ameliorating human and environmental health. In the past few years, the government has recognized the importance of promoting the sustainable management of India's water resources and has placed water development as one of its main priorities in the coming decades.

Water pollution through the discharge of waste water from industrial sites can cause vast environmental problems. Water used in the cooling process is often discharged from the plant at a high temperature, if this water then goes straight into the waste systems or worse, into a river system, the temperature can cause significant environmental damage including the destruction of plant and animal species that naturally survive in cold water and the reduction in the overall water quality of the area. Temperature is not the only factor in the issues of waste water, if the water is discharged from the process rather than the cooling; there is the chance that it is polluted in some way.

Businesses can counteract the problems of waste water by re-using as much as they possibly can. Waste water from cooling and waste water from processing can both be filtered and used for WC's or garden watering etc. in addition to rainwater which can be collected for use in similar ways.

In order to address the problems of waste water, companies need to invest in measures to reduce the amount of water used in their public conveniences, clean their discharge before it enters the natural systems and utilise "grey" water as much as possible. Such requirements are already in force because of the Rio Summit, companies have a responsibility for the environment in which they are situated, and this is iterated in the guiding principles. If they fail to implement means to reduce their pollution they are not only liable for the costs of any clean up under the "Polluter Pays" principle, but they are also bound by the "precautionary principle" which emphasizes the need to be cautious and ensure an acceptable level of care is taken.

Natural Resource Conservation

As stewards of the land, we must protect the quality of our environment and conserve the natural resources that sustain it by

implementing conservation practices that improve water quality, conserve water and energy, prevent soil erosion and reduce the use of nutrients and pesticides.

Agricultural Industry Sustainability

Educating the next generation of farmers, welcoming newcomers and ensuring the safety and well-being of today's farm workers are critical components of maintaining a profitable, strong agricultural industry poised for a bright future.

Annextures:

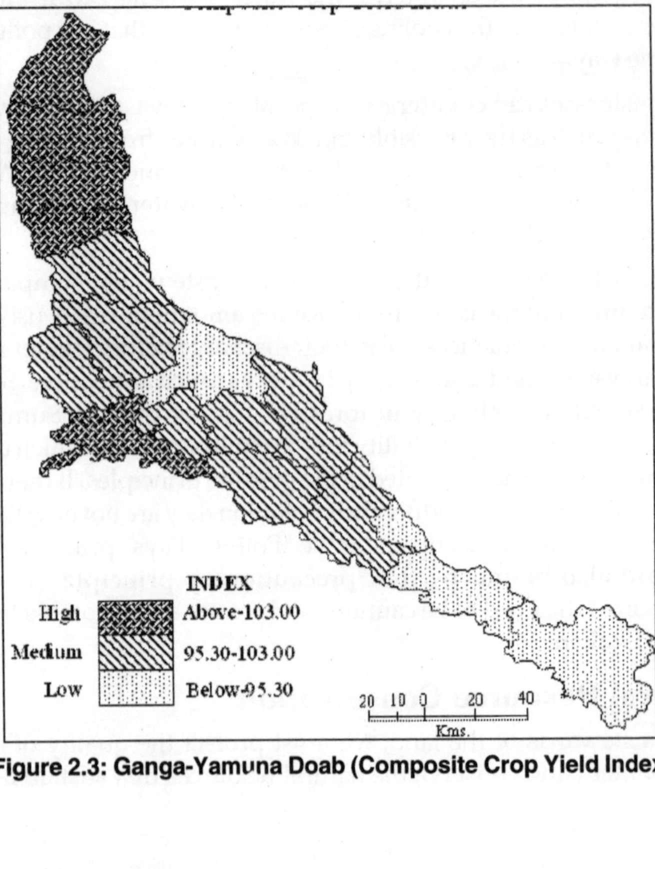

Figure 2.3: Ganga-Yamuna Doab (Composite Crop Yield Index)

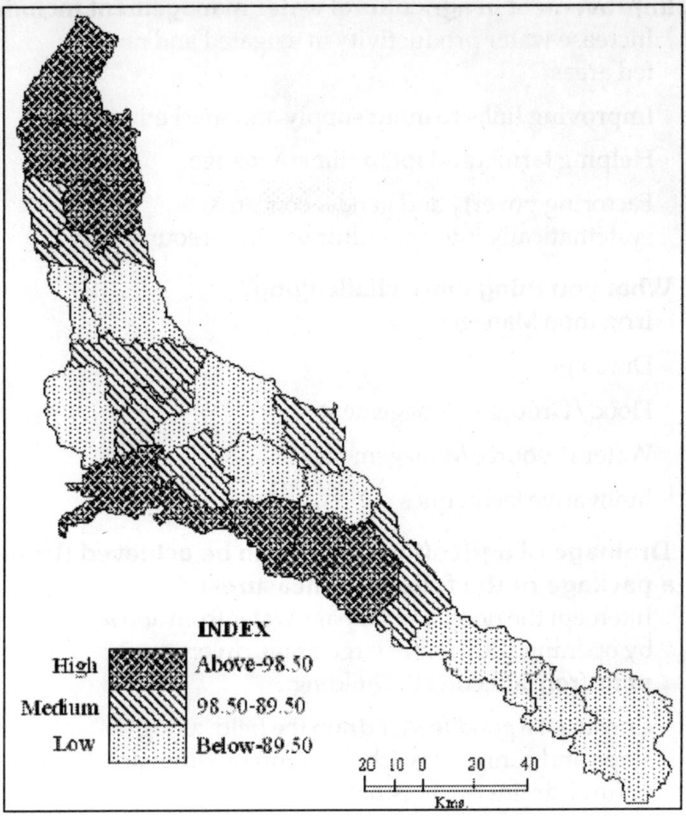

Figure 2.4: Ganga-Yamuna Doab (Composite Crop Yield Index 2000–04)

Questionnaire

Q1. Do you understand water resource management?

Yes ❑

No ❑

Q2. What challenges are there in water resource management?

Economic Development ❑

Population Growth ❑

Industrial Development ❑

Climate Change ❑

Q3. Improvement in agricultural water management includes:

- Increase water productivity in irrigated and rain fed areas ❑
- Improving links to input supply and marketing chains ❑
- Helping farmers adapt to climate change ❑
- Factoring poverty and gender concerns systematically into agricultural water resources ❑

Q4. What you thing more challenging?

- Irrigation Management ❑
- Drainage ❑
- Flood/Drought Management ❑
- Water Resource Management ❑
- Innovative techniques ❑

Q5. Drainage of agricultural lands can be achieved through a package of the following measures:

- Intercept the flood and seepage water from above by opening sufficiently large drain, divert the same from affecting the holding ❑
- Construct a good feeder drain the field, along the slope and connect it with the common drain or natural drains such as nallas ❑
- Construct adequate drains in the plots and crop fields and connect them to the vertical disposal drain ❑
- Provide drop pits and stone pitching in erodible spots ❑

Q6. Successful reclamation and management of salt affected soils requires?

- Proper Diagnosis of the Nature ❑
- Intensity of the problem ❑
- Suitable measures adopted for its control under given resource conditions ❑

Q7. What do you understand by agriculture development?

Agriculture development brings about a revolution in the agriculture industry to give birth to an agriculture which is profit giving and at the same time eco friendly ❑

Agriculture development means providing assistance to the crop producers with the help of various agricultural resources ❑

Technical help for land use is also an important part of agriculture development ❑

Q8. Land issues relating to agricultural land include:

Preservation of resources associated with agricultural land ❑

Reducing disputes between the agricultural community and the residents in the rural area ❑

Agriculture development also includes helping out farmers with regard to animal agriculture ❑

Q9. Do you think there is a scope of agriculture productivity in Ganga Yamuna plain?

Yes ❑

No ❑

Q10. What are the sources of water?

Under River Flow ❑

Ground Water ❑

Desalination ❑

Frozen Water ❑

References

1. agro.indiamart.com/agriculture–products/
2. ddinajpur.nic.in/flood management/flood_management.html
3. dnr.wi.gov/environment protect/water.html
4. en.wikipedia.org/wiki/Ganges
5. en.wikipedia.org/wiki/Irrigation management
6. en.wikipedia.org/wiki/Water resources

7. en.wikipedia.org/wiki/Water resources

8. en.wikipedia.org/wiki/Water resources

9. *en.wikipedia.org/wiki/Yamuna –*

10. http://en.wikipedia.org/wiki/Watershed_management

11. https://www.soils.org/story/2009/.../crop–water–use–efficiency

12. Munir, A., 1988. Agricultural productivity and regional development: A case study of the sub-Himaliyan east region of Uttar Pradesh. *The Geographer*, 35(2): 48–49.

13. planningcommission.gov.in/reports/sereport/ser/ser_ap.pdf

14. Rehman, H., 1976. *Mechanization of Farming and its Impact on Food Crop Productivity in Uttar Pradesh.*

15. Rehman, H., 2003. *Spatial Distribution of Agricultural Productivity and Its Correlates in North Bihar Plain.*

16. *The Geographer*, 23(2): 43-56, July.

17. *The Geographer*, 50(1): 73-84, Jan.

18. web.worldbank.org › Topics › Water

19. www.alibaba.com/showroom/agriculture-prod.html

20. *www.derm.qld.gov.au/environmental_management/water/environmental...environmental_protection_water.../index.html -*

21. www.gisdevelopment.net › Geospatial Application Papers

22. www.nri.org/.../WHIRL per cent 20working per cent 20paper per cent 207_final.pdf

23. www.slashbuster.com/drainage-management.htm

24. www.unescap.org/.../flood per cent 20management per cent 20in per cent 20ganga.doc

25. www.water.ca.gov/wateruseefficiency/

Chapter 3

Water for Agriculture, Food Security and Livelihood

☆ *Abdul Khan*

ABSTRACT

Population growth and climate change are two main challenges for our food security. World projected population will be approx. 9.00 billion in 2050 against the present population of approx. 6.5 billion. Climate change is the other factor which will affect our water resources due to change in rainfall pattern and its intensity in time and space, temperature rise, wind movement and humidity. There are two types of agriculture *viz.* irrigated and rain-fed agriculture. The crop yield in irrigated area is 2 to 3 times more than the rain-fed area. At present, there are approx.1500 million hectares land under cultivation, in which, only 277 million hectares land *i.e.*18 per cent area is under irrigation which provides 40 per cent crop yield, remaining 82 per cent area which is rain-fed provides 60 per cent crop yield. Conservation and management of water resources in irrigated and rain-fed area is essential for development of high yield agricultural production. Conservation of soil and water resources along with the development of high yield variety seeds are key factors for increase in agricultural production. For food security, the maximum agricultural production per unit area of arable land with the optimum utilization of the available water is required to be achieved that is only possible with the integrated natural resources development and management. Diversification of agriculture towards horticulture (production of fruits, vegetable, flower, aromatic and medicinal plants etc.); animal husbandry (rearing of cow, buffalo, sheep and goat etc.); diary; poultry; fishery and agro-forestry will provide new opportunities of livelihood and income generation. Agro forestry will also reduce the impact of climate change.

Keywords: *Population growth, Climate change, Irrigated and rain-fed agriculture, Food security, Livelihood.*

Water for Agriculture: An Introduction

India gets approximately 4000 BCM (billion cubic meters) of water every year from all natural sources. Out of this nearly 700 BCM are lost in evaporation and other nearly 700 BCM are lost during the flow on ground and a large part of water nearly 1500 BCM flows into the sea as floods. Thus, the remaining only 1120 BCM is available (690 BCM as surface water and 430 BCM as ground water) for our needs. Approximately 70 per cent of this water is being utilized in irrigation sector. The geographical area of India is approx.329.00 mha, in which arable land is 166.00 mha. Our ultimate irrigation potential is 139.893 mha. We have created an irrigation potential of 93.946 mha and our utilizing irrigation potential is 87.26 mha. Our net irrigated area is 62.29 mha against the net sown area of 140.86 mha. Our gross irrigated area is 87.26 mha (in which 34.97 mha is double irrigated area), against the gross sown area of 195.83 mha (in which 54.97 mha is double sown area).

There are two type of agriculture: *viz.* irrigated agriculture and rain-fed agriculture.

Water Management in Irrigated Agriculture

We have to optimum utilized the water of the irrigated system for the maximum production of the crop yield; it can be achieved in the following manner;

1. Reducing the conveyance losses
2. Changing the irrigation pattern
3. Changing the cropping pattern

4. Adopting participatory irrigation management
5. By formation of water users association

Management of Water in Rain-fed Agriculture

1. Approximately 60 per cent area of agriculture is rain-fed in India.
2. Rainfall is the only source of water in rain-fed farming.
3. To maintain the soil moisture for the growth of crops the ON FARM and OFF FARM rain water conservation and management is very essential for supplemental irrigation.
4. ON FARM rain water harvesting and management may be achieved by the following methods:
   (*a*) Open wells
   (*b*) Closed wells/tube wells
   (*c*) Open ponds/tanks
   (*d*) Underground storage tanks
   (*e*) Use of sprinkler/drip irrigation
   (*f*) Watershed based water harvesting
   (*g*) Recharging of ground water.
5. OFF FARM rain water harvesting may be done by the following methods:
   (*a*) Gully plugging
   (*b*) Loose stone bundhs
   (*c*) Check dams
   (*d*) Gabions
   (*e*) Anicuts
   (*f*) Spillways
   (*g*) Earthen bundhs
   (*h*) Runoff storage structures.
6. OFF FARM water conservation may be achieved by drainage development:

   The rainfall water of any catchment area finally reaches to its drainage line. This water may be collected for its future need in off time of rainfall duration by the construction of a series of water retaining structures in the path of each drainage system. These structures should be constructed in such a fashion that the water may not flow above the top line of the drain. The water may be collected in the bed of the drain from head to tail during the monsoon season and this water may be utilized in off raining season. This water may also recharge the ground water and may be lifted for irrigation purposes in hour of needs.

7. These structures may be constructed in a series at such a interval from head to tail, according to slope of drain that whole drain looks like a mini reservoir.
8. All the drainage system as a whole may be utilized as a water body to harvest the rain water and reserve it for future needs.
9. These structures may be weir or barrage depending upon the nature of silt load.
10. These structures may be submerged or non-submerged according to the discharge of the drain.
11. These structures may be temporary or permanent type as per local need of people.
12. All these harvested water in monsoon season may be utilized in non-monsoon season for irrigation of crops.

Water Management in Himalayan Region

Mountain people of Himalayan region are very resource poor due to remote environment, degraded infrastructure, socio-economic backwardness, and low land holdings. Their main source of livelihood is agriculture which is mostly rain-fed that entirely depends on mercy of rains. Rain fall is very uneven in time and space and they face the vagary of rain fall in the form of low yield. Due to lack of irrigation facilities, their farming is hardly sustenance in nature, poverty is prevailing there and migration is very common.

Himalayan region is very rich in water resources. All the major rivers of northern India originate from Himalaya. This area is water surplus in nature but due to the difficult topography and terrain of region, we are lacking behind in the development of irrigation systems. We can develop the irrigation system in the following ways:

1. *Contour irrigation*: This is very common in hilly areas near the natural streams; we make a diversion head and divert the water in channels as gravity flow.
2. *Lift irrigation*: This is not very common in hilly areas but it constructed in those areas, where other type are schemes are not feasible. It requires external energy for lifting the water.
3. *Tank irrigation*: This is a very popular form of irrigation in hilly regions but it is suitable only for small areas. It is suitable in upper reaches where rain fall is the only source of water.
4. *Sprinkler irrigation*: This is a water saving scheme where there is scarcity of water. Its initial cost is high.
5. *Drip irrigation*: This system is suitable for orchards where there is water scarcity.
6. *Hydram irrigation*: This system is also a short of lift irrigation for small areas; it requires no any external energy for lifting the water.
7. *Rejuvenating the springs*: By constructing a series of check dams in the path of natural streams/rivulets, we can recharge the ground water with the retention and detention of flow of water and this recharge water will flow as sub-surface water and will intensify the discharge of springs and springs will rejuvenate. These springs will help us in irrigation of crops and will also solve other needs of water. These check dams will silt up the streams/rivulets that will improve the stability of slopes that is a safety against land slide.
8. *Sub-surface dyke cum check dams*: In Himalayan foot hills region, there are a large number of natural streams which are seasonal in nature but they have sub-surface water. By construction of sub surface dyke cum check dams at suitable places in natural streams, we can check the sub-surface of flow of water and conserve this water for surface irrigation and recharge the ground water for tube well irrigation in the foot hills of Himalayan region.
9. *Artisan wells*: In the Himalayan foot hills region, there is a great scope for artisan well, we can investigate and develop the artisan wells in foot hills region.

10. *Reservoirs*: It will help us in irrigating the agriculture in non-monsoon seasons and lean period of monsoons.

Food Security

Various studies by demographers in India and abroad suggest that world population will continue to rise at least till 2050 AD. This indicates India's 2050 AD population as 1349 million (low projection), 1640 million (medium projection) and 1980 million (high projection). In India the food grain availability is at present around 525 gms per capita per day. Assuming the same level of consumption, which although is supposed to rise with improvement in economy and resultant higher standard of living, the annual food grain requirement will be about 315 MT. If small raise is made in per capita consumption to 650 gms, the food grain requirement will be about 390 MT. Taking the projection of about 1800 million by 2050 AD as reasonable, it would require about 430 MT of food grain annually at the present level of consumption.

The present productivity of irrigated land is about 2.5 T/ha and less than 0.5 T/ha for rain fed lands. Assuming that these levels can go up to 3.5 and 1.0 T/ha respectively by 2050 AD, it is imperative that we create an irrigation potential of at least 130 m.ha for food crop alone and 160 m.ha for all crops to be able to meet demands of the country by 2050 AD. This will mean a much greater use of available inputs. First input is water, the second input would be fertilizers and third input would be much larger emphasis on research. But even if it is taken for granted that no "Wonder" food grain seed will be available, we will have to increase the area under irrigation from present 28 per cent to roughly about 40 per cent by the year 2050. The possible options available to meet the future irrigation and food needs may, therefore include drip and sprinkler irrigation, change of cropping pattern, evaporation control and reduction of conveyance losses, recycling of water, inter-basin transfer, de-salinization of sea water.

Livelihood

Diversification of agriculture will help the rural people in creation of employment opportunities and income generation that will help the rural mass for their poverty alleviation. The following activities may be adopted in the diversification of agriculture.

1. *Horticulture*: production of fruits, vegetables, flowers, aromatic and medicinal plants.
2. *Animal husbandry*: rearing of cows, buffaloes, sheep, goats etc. Development of dairy farm, poultry farm etc.
3. *Fish farming*: development of fisheries farms etc.
4. Agro-forestry and Agro-industries

The Impact of Climate Change

The impact of climate change will affect our water resources due to change in rain fall pattern in time and space. Rise in temperature means more evaporation losses, change in wind movement and humidity. That will require more water for our agricultural production. So, we have to conserve more water in form of reservoirs or dams for our need in water scarcity period.

Conclusion

The whole paper can be summarised as below:

- ☆ Optimum utilization of the irrigation potential created *i.e.* 93.946 mha.
- ☆ Developing the ultimate irrigation potential *i.e.* 139.893 mha.
- ☆ Harnessing the 1500.00 BCM of flood water which is not utilized so far.
- ☆ Protecting the 40.00 mha of flood affected areas.
- ☆ Management of the rain-fed areas as per watershed development scheme.
- ☆ Integrated natural resources development in an environmental friendly way.
- ☆ We have to produce 400.00 MT of grains to feed our projected population of 1800.00 million of 2050 AD.

Chapter 4

Impact of Climate Change on Snow Melt Runoff: A Case Study From Langtang Basin Nepal

☆ ***Bikas Bhattarai and D. Regmi***

ABSTRACT

Observations and model-based studies suggest that the substantial hydrological changes in mountain watersheds where hydrology is dominated by cryospheric processes (IPCC 2007). The response of cryospheric processes to a warming climate in mountainous areas can be analyzed by examining the responses in the seasonal and annual hydrologic regimes of rivers where snow and ice contribute significantly to the runoff. This study which aims to access the impact of potential warming on river discharge utilizes a snowmelt runoff model in the Langtang River Basin in Nepalese Himalaya. For this data were derived from remotely sensed snow cover from Moderate Resolution Imaging Spectroradiometer (MODIS). The Snowmelt Runoff Model (SRM) is calibrated using daily stream flow from 2005 to 2006 and the stream flow can be predicted with a high degree of accuracy. Although snowmelt hydrology has been examined at a large watershed scale in the Hindu Kush-Himalaya region, changes in hydrologic processes with climate change need to be examined and compared across finer basin scales for assessing water availability and vulnerability.

Three climate change scenarios were used to drive the model in order to understand the impact of changing conditions. A scenario of a 1°C temperature increase and results in an increased runoff volume by ~5 per cent, with stream flow exceeding present conditions in all months. Similarly temperature is increased 2 and 3°C to the present condition in such case discharge increase in 10 per cent and 16 per cent by volume annually. These results show the significant impact of increasing temperature on snowmelt runoff and importance of improved monitoring and modeling to better understand the impact of climate change on hydrology.

Keywords: *Climate change, Himalayas, Langtang basin, MODIS snow, Snow Melt Runoff (SRM).*

Introduction

Growing evidence indicates that the low-latitude mountainous cryosphere is undergoing to change due to recent warming (IPCC 2007). Geographic areas where the water cycle is dominated by snowmelt hydrology are expected to be more susceptible to climate change as it affects the seasonality of runoff (Adam *et al.*, 2009). Changes in seasonal snow cover and glaciated regions may alter the variability of stream flow and hence water availability which sustains a large population downstream. Despite its regional importance, there is uncertainty associated with the rates and magnitude of climate change impacts on snow cover and snowmelt hydrology. These climate-driven responses of mountainous river hydrology when combined with potential land cover changes, population growth, and already stressed water resources may pose significant challenges for this region.

Since the negative impacts of climate change to water resources are expected to be greater than its benefits (Kundzewicz *et al.*, 2007), it is important to understand the hydrologic variability particularly in mountain environments with respect to potential climatic changes. Regional climate projections by the IPCC (2007) indicate that Central Asia to be warmed by a median temperature of 3.7°C by the end of the 21st century, with largest warming over higher altitudes particularly the Tibetan Plateau and the Himalayas. According to the IPCC, the temperature increase in the Himalayan region has been greater than the global average of 0.74°C over the last 100 years.

Several global scales and few local to regional scale hydrological studies have been conducted to assess impacts of a changing climate on snowmelt hydrology. Hydrological simulations of large, continental river basins to climate change in 2025 and 2045 assessed through global climate models showed that snow-dominated basins of mid to higher latitudes showed the largest changes in the hydrological cycle (Nijssen *et al.*, 2001). In these scenarios, although the coldest snow dominated basins exhibit an increase in the spring stream flow peak, the basins with more transitional snow regimes exhibit a reduced snow melt peak and increased winter stream flow. Analytical studies representing temperature increases of 1-3°C in the western Himalayan region suggest an increase in glacial melt runoff by 16-50 per cent (Singh and Kumar 1997), regression of the annual maximum spring stream flow period by a month, and increased sensitivity of snow-fed basins in terms of reduction in water availability (Singh and Bengtsson 2005). Water resource management and the evaluation of impacts of climatic change require quantification of stream flow variability and hydrologic models provide a framework to investigate these relationships (Leavesley 1994). Poor accessibility and adequate network of hydro meteorological station in high altitude regions is a major impediment to runoff modeling and as a result only a few studies have explored snowmelt runoff models in Himalayan sub basins (Akhtar *et al.*, 2008; Braun *et al.*, 1993; Shrestha *et al.*, 2004). The impact of potential climate change on snowmelt hydrology in the Himalaya, therefore, remains an active area of research. The objective of this paper is to assess the impact of potential warming on river discharge. Based on the future projected climatic scenarios for the Himalayan region (IPCC 2007), three climate scenarios where temperature is increased, were used to assess and quantify the potential changes and effects of a warmer climate on snowmelt runoff and total discharge.

Study Area

Langtang River basin is situated at about 60 km north of Kathmandu at the border to Tibet, and is the head water area of Trisuli River in the Narayani River System. The total basin area of Langtang River is 363 sq.km. The catchment reaches from the 3800 masl. up to the Langtang Lirung at 7234 m ASL. with two other peaks above 7000 m ASL (Konz *et al.*, 2005). The average altitude is 5169 meter above sea level (masl) with mean slope of 26.7° which

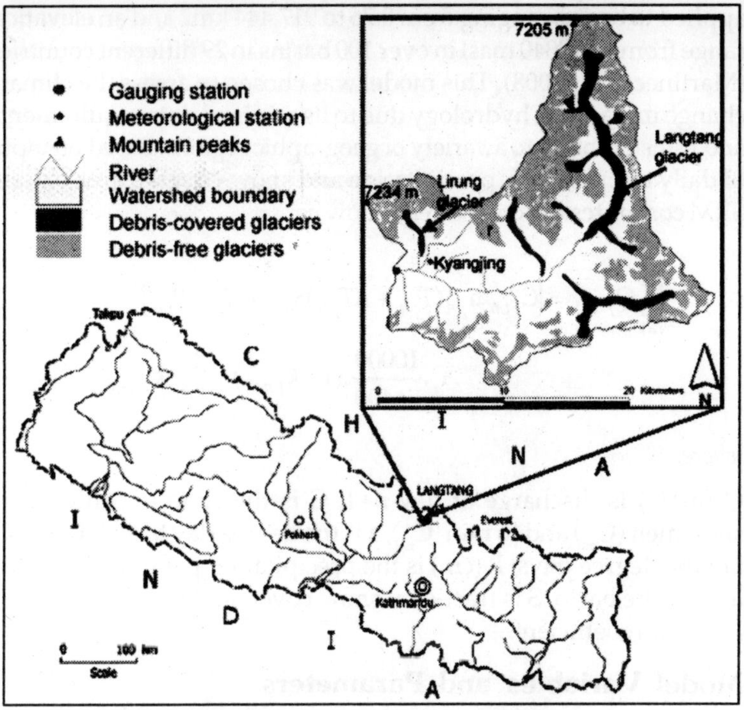

Figure 4.1: Location of Study Area

reflects the high potential relief energy of the catchment. Glacier covers an area of 166 km² of the catchment, of which 32 km² are covered by debris, especially the glacier tongues below 5200 m (Burgi *et al.,* 2009). At the meteorological station the mean daily air temperature for the hydrological years 01/02 to 05/06 is -0.7 °C in the dry season from October to June and 6.9 °C in the Monsoon season. The mean annual temperature, with the corrected time series is 1.81°C in the years (DHM 2066).

Model Structure

The modeling of the stream discharge was based on the Snowmelt Runoffs Model (SRM) developed by Martinec (2007). SRM is a conceptual, deterministic, degree day hydrologic model that simulates and forecasts daily runoff resulting from snowmelt and precipitation from hydro-meteorological input data. SRM has been applied to basins ranging from 0.76 to 917,444 km2 and an elevation range from 0 to 8840 masl in over 100 basins in 29 different countries (Martinec *et al.*, 2008). This model was chosen to assess the climate change impacts on hydrology due to its minimal data requirements and transferability to a variety of geographic regions. Based on input of daily temperature, precipitation and snow-covered area values, SRM computes the daily stream flow as:

$$\sum_{j=1}^{m} Q_{j.n+1} = \left[C_{Sj.n} .a_{j.n}(T_{j.n} + \Delta T_{j.n})S_{j.n} + C_{Rj.n}P_{j.n}\right]$$

$$A_j . \frac{10000}{86400} .(1 - k_{j.n+1}) + Q_{j.n}k_{j.n+1} \quad (1)$$

where,

Q (m^3/s) is discharge at day n+1, C is the runoff coefficient to snowmelt (C_{sn}) and to rain (C_{rn}), a is the degree day factor, T+ΔT (C) are the degree days, P (cm) is the precipitation per day, A (km^2) is area of the basin, S is fractional snow cover, and k is the discharge recession coefficient.

Model Variables and Parameters

Basin Characteristics

Zonal Characteristic

The area-elevation curve is used in SRM to determine the zonal mean hypsometric elevation (*i.e.* elevation dividing the area in to equal half), and the value of the hypsometric elevation is then used as the elevation to which base station temperatures are extrapolated for the calculation of zonal degree days. This area-elevation curve and zonal mean hypsometric elevation can also be derived from DEM by using the spatial analysis capabilities provided in a Geographical Information Systems (GIS). For the Langtang River basin, the basin and elevation zone areas, elevation ranges, and mean hypsometric elevations for each of nine zones are computed and listed in Table 4.1 and elevation zones are delineated in intervals of 400 m.

Model Variables

Temperature and Number of Degree-Day

In SRM the number of degree-day is expressed in the form of temperature. Temperature is a comprehensive index for calculating

Table 4.1: Elevation Zone, Zonal Area, Elevation Range, and Mean Elevation for the Nine Elevation Zones of Langtang Khola Basin

Elevation Zone	*Elevation Range/m a.s.l.*	*Elevation Area/km²*	*Zonal/Basin Area (per cent)*	*Hypsometric Elevation of Zone/m a.s.l.*
A	3600-4000	18.12245	4.9812	3933
B	4000-4400	33.978	9.3393	4246
C	4400-4800	51.8438	14.25	4655
D	4800-5200	84.8030	23.30	5005
E	5200-5600	78.7392	21.642	5385
F	5600-6000	61.6264	16.939	5766
G	6000-6400	25.4029	6.9823	6141
H	6400-6800	7.7946	2.1425	6575
I	6800-7200	1.5746	0.4144	6861
Total		363.86	100 per cent	

snowmelt, moreover, temperature is a routine meteorological observation and easy to acquire, extrapolate and forecast. In order to compute the daily snowmelt depth, the average number of degree-day is computed in each zone as:

$$\overline{T} = \frac{T_{max} + T_{min}}{2} \quad (2)$$

where,

$\overline{T}$ is average number of degree-day; T_{max} and T_{min} are daily highest and lowest temperatures, respectively. The adjustment value of number of degree day for each elevation zone is computed as:

$$\Delta T = \gamma.(h_{st} - \overline{h}).\frac{1}{100} \quad (3)$$

where,

γ is temperature lapse rate; h_{st} is altitude of the temperature base station; $\overline{h}$ is mean hypsometric elevation for a given zone. The lapse late in this research is taken as 0.53 °C/100m (Kayastha *et al.*, 2005).

Precipitation

During the snowmelt season, precipitation usually occurs in two forms, rain or snow. A critical temperature in SRM is adopted to decide whether a precipitation event will be as rain or snow. When temperature is higher than critical temperature, the precipitation is determined to be rain; otherwise, the precipitation is determined to be snow. The precipitation increases with altitude is usually true in the snow and glacier mountain areas in Nepali Himalayas too (Higuchi *et al.*, 1982).

There is 1.3 times greater precipitation at 5000 m than that at elevation 4000m (Seko, 1987). The spatial distribution of the precipitation was made as the equation given by the Seko, (1987).

Snow Covered Area

It is a typical feature of mountain basins that the areal extent of snow cover gradually decreases as the snowmelt season progresses. The snow coverage is a very important input variable to SRM, and daily snow coverage values can be obtained from the depletion curves, while the depletion curves of the snow coverage can be

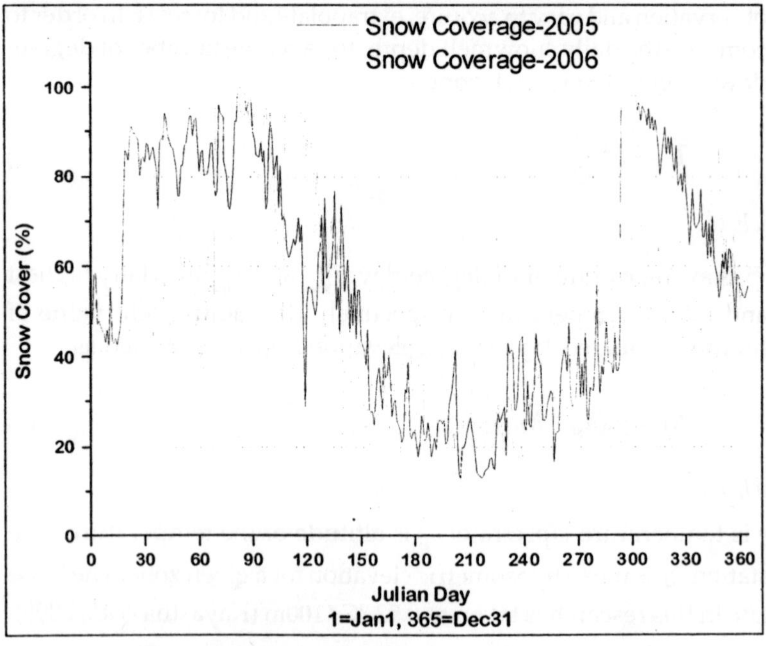

Figure 4.2: Snow Coverage in the Study Area

prepared from periodical snow cover mapping. First, elevation zones with an equidistance of 400 m are derived from basin's DEM, resulting in a map layer of elevation zones. Step two overlay the map layer of elevation zones on the map layer of snow cover. The result of this process is the snow coverage per elevation zone. Using a time series of snow coverage data, the temporal and spatial variations of snow coverage are derived and plotted.

Model Parameters

Runoff Coefficient

The SRM accepts separate values of runoff coefficient for snow and rain, because runoff coefficient is usually different for snowmelt and for rainfall. For the Langtang River basin, the runoff coefficients of snow and rainfall are determined from the basin characteristics. The adjustment of runoff coefficient is always required in the initial stage of model simulation, especially when a runoff simulation is not at once successful. In this study the runoff coefficient is treated as the parameter to be calibrated.

Degree-day Factor

The degree-day factor in SRM is a key parameter for calculating snowmelt expressed in depth of water:

$$M = a.T \tag{4}$$

where,

M is snowmelt in depth of water (cm); "a" is degree-day factor; and T is number of degree-days. The degree day factors for snow and ice ablation used in this research are 7.0 and 8.0 mm $d^{-1}{}^{\circ}C^{-1}$ respectively at the altitude up to 5000 m a.s.l. And above 5000 m a.s.l the factors are 10.5 and 9.5 mm $d^{-1}{}^{\circ}C^{-1}$ was used respectively (Kayastha *et al.*, 2005).

Critical Temperature

The critical temperature is an average value of temperature used in SRM to determine whether the precipitation is rain or snow. Critical temperature is usually higher than freezing point and diminishes to be close to 0°C as snowmelt season progresses. The estimation of critical temperature for determination of snowfall amount during precipitation events is carried out using the relation obtained on glacier AX010: east Nepal. The relation was obtained by plotting calculated monthly mean air temperature (Kayastha *et al.*, 2005).

Rainfall Contributing Area

Rainfall contributing area is a parameter used in SRM to determine whether rainfall runoff is added to snowmelt runoff only from snow-free area (option 0) or from entire basin or zone area (option 1). If option 0 is set in SRM simulation, rainfall runoff is added to snowmelt runoff only from the snow-free area; otherwise if option 1 is set, rainfall runoff is added to snowmelt runoff from the entire basin or zone area. For simplicity, the option 1 is assumed in this research. The melting effect of rain is neglected because the additional heat supplied by the liquid precipitation is considered to be small (Wilson, 1941).

The Recession Coefficient

It is an important model parameter since (1-*k*) is the proportion of the daily melt-water of snow contributing to the daily runoff. Recession coefficient can be obtained by the analysis of historical discharge data according to the equation:

$$k_{n+1} = x.Q_n^{-y} \qquad (5)$$

where,

Q_n is discharge on day n, x and y are two constant. For the determination of x and y, daily discharge on a given day, Q_n is plotted against the value on the following day, Q_{n-1}. As illustrated in Figure 4.3, the Langtang River basin in 2006, the lower envelop line of all points is considered to indicate the *k* value. Based on the relation $k = \frac{Q_{n+1}}{Q_n}$, it can be derived that k_1=0.937 for $Q_1 = 2.47\frac{m^3}{s}$ and k_2=0.723 for $Q_2 = 9.12\frac{m^3}{s}$. According to the equation (12) it follows, *x*=1.146 and *y*=0.2082, the recession equation for the Langtang Khola using the lower envelops line thus becomes:

$$k_{n+1} = 1.146.Q_n^{-0.2082} \qquad (6)$$

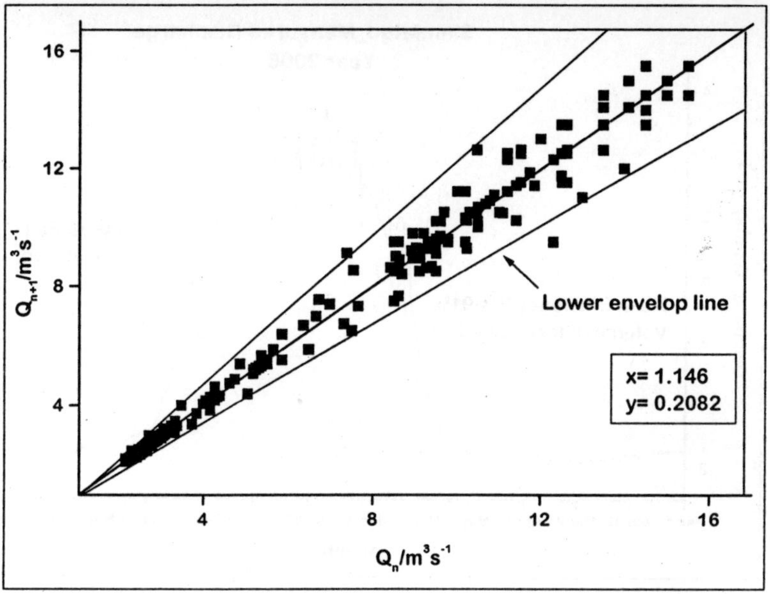

Figure 4.3: Recession Flow Plot Q_n vs. Q_{n+1} for Langtang Basin

Snow Melt Runoff Simulation

Since the 2006 is the average year and have good quality of data, so this year is taken as base year for both simulation and investigate the impact of climate change while validation is done in 2005. The different variables and parameters of each zone are loaded as input in SRM model. Then the model is run for the calibration and simulation. The simulated hydrograph is shown in the Figure 4.4.

The measured discharge is well simulated in the year 2006 with coefficient of determination (R^2) 0.91 (Nash efficiency) and volume difference (ΔV) less than 0.1 per cent. Few stream flow peals could not be generated in simulated runoff, which was possibly due to sudden release of stored water at some location in the glacier body, and such events are clearly identified because they are not supported by climate condition (Singh *et al.*, 2008).

In the validation year 2005, there is good coefficient of determination of (R^2) 82 per cent and volume difference (ΔV) 5.35 per cent (Figure 4.5). The discrepancy in peal flow is due to reason

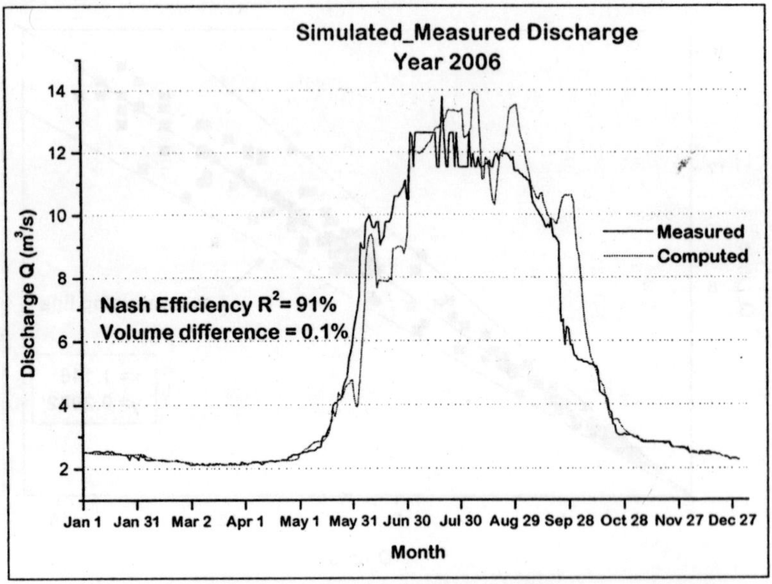

Figure 4.4: Simulated and Measured Discharge in the Year 2006

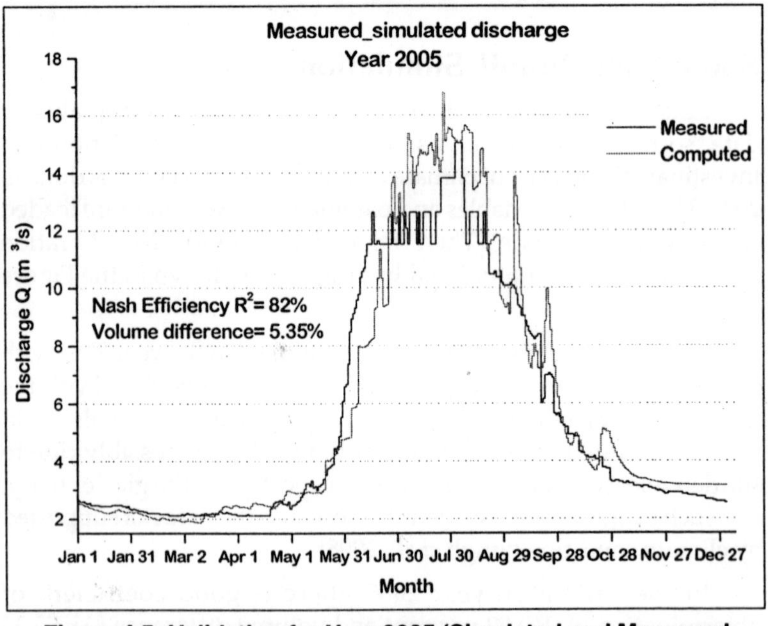

Figure 4.5: Validation for Year 2005 (Simulated and Measured Discharge)

as explained earlier. But considering volume difference is due to the following reason

1. The measured discharge in the year 2005 is 173.855 Mm^3 which is more than in the year 2006 (*i.e.* 171.265 M m^3).
2. Precipitation in the base year (2006) is 10.36 per cent less than in validation year *i.e.* 2005.

Contribution of Snowmelt in Stream Flow

The contribution of snowmelt in the stream flow is still considerable in the stream flow though there is high precipitation (334.56 M m^3) than the runoff (171.26 M m^3).

There is 31.55 per cent contribution by snowmelt in winter season and 17.75 per cent in summer season altogether 19.46 per cent contribution in the annual stream flow runoff (Figure 4.6) in the year 2006. The low snowmelt in the January is because of the low temperature recorded during that period, which shows average monthly temperature is -0.37°C respectively in that period.

Similarly in year 2005 the contribution of snowmelt in stream flow is 23.94 per cent in the annual flow with 36.60 per cent

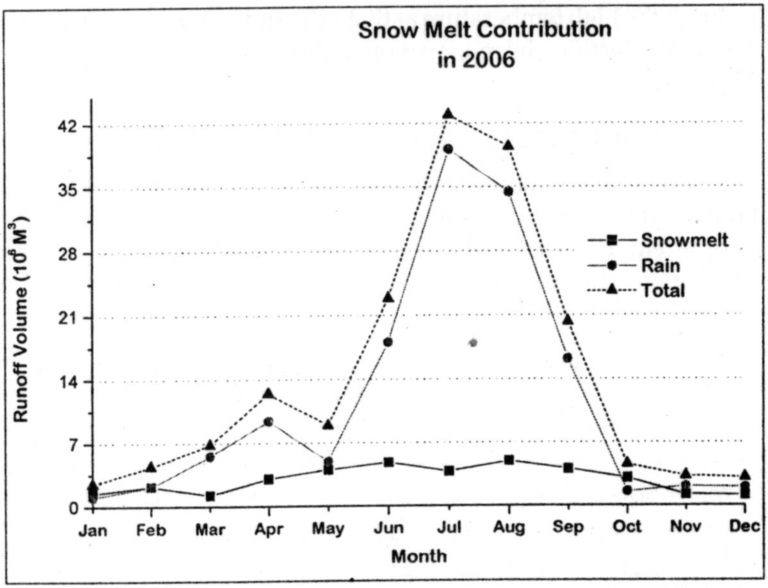

Figure 4.6: Contribution of Snow Melt in Stream Flow in 2006

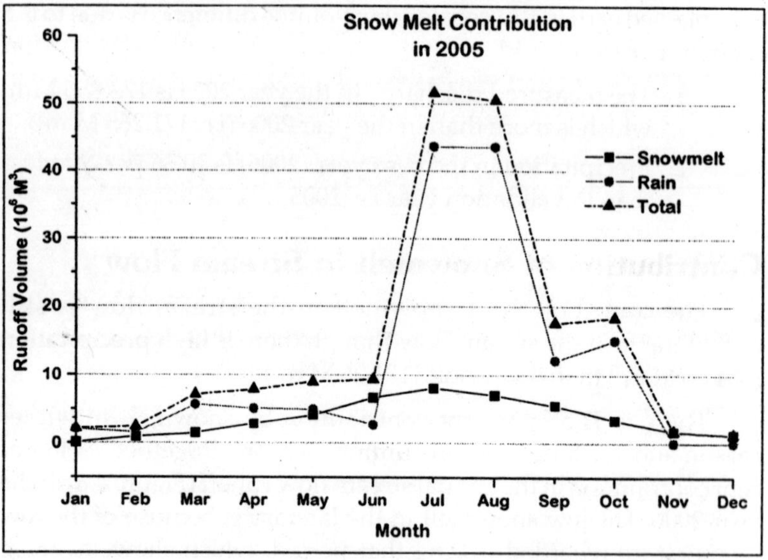

Figure 4.7: Contribution of Snowmelt in Stream Flow in 2005

contribution in winter season and 24.47 per cent contribution in summer season (Figure 4.7). The increase contribution of snow melt is due to the high temperature at that year and more snow coverage. This is also justified by more runoff in this year 2005 than in the year 2006.

Impact of Climate Change

After the simulation has been done in the year 2006, this year is taken as the base year to investigate the impact of climate change (rising temperature) in the snow melt runoff. In climate change simulation in SRM, only the temperature is changed from the ΔT1, 2 and 3 °C because there is no definite trend of change in precipitation.

Table 4.2: Impact of Climate Change in Contribution of Snowmelt in Stream Flow

Season/Period	*Present*	*T+1ºC*	*T+2ºC*	*T+3ºC*
Winter Season	35 per cent	37 per cent	38 per cent	44 per cent
Summer Season	18 per cent	23 per cent	27 per cent	32 per cent
Annual	19 per cent	23 per cent	28 per cent	33 per cent

From April to Dec, base year 2006

The impact of climate change (*i.e.* temperature) is increasing trend in stream flow with increased snow melt contribution approximately at the rate of 4 per cent in the annual and 2 per cent in winter flow per one degree temperature rise.

Table 4.3: Impact of Climate Change on Runoff Volume

Season	*Meas-sured*	*Present*	*T+1ºC*		*T+2ºC*		*T+3ºC*	
	M m3	*M m3*	*M m3*	*Increase in%*	*M m3*	*Increase in%*	*M m3*	*Increase in%*
Winter	34.24	34.68	35.20	2%	36.31	4%	37.74	8%
Summer	137.01	138.8	147.0	6%	155.6	11%	167.9	17%
Annual	171.26	173.48	182.2	5%	192.0	10%	205.6	16%

This pattern is shown in the Table 4.2 and Figure 4.8. The main factor of such increase in snowmelt contribution is melting of snow and glacier in the basins since there is 71 per cent area of the total watershed above 4800 masl where annual average temperature is less than 0 ºC and the snow coverage is more than 63 per cent average

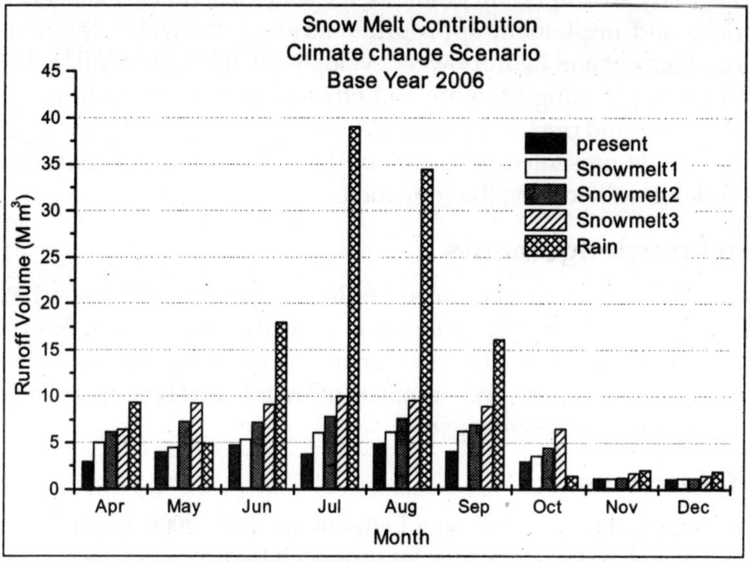

Figure 4.8: Impact of Climate Change on Contribution of Snowmelt in Stream Flow

in year. As a result, though the temperature is raised up to 3 °C, there will be enhanced melting in the melting period while balanced snow deposition in the accumulation period in the winter. That's why the result of impact of climate change is as anticipative.

Conclusion

The result shows that the Positive Degree Day or Temperature Index method coupled in Snowmelt Runoff Model, SRM can be well applied in rugged and remote Himalayan catchment like Langtang Khola basin with limited data. The study area is glacierized basin with more than 50 per cent annual snow coverage. The contribution of snowmelt in stream flow was found as 33.98 per cent in winter, 20.84 per cent in summer and 21.58 per cent in annual. The study demonstrates that the impact of climate change (*i.e.* temperature) to stream flow is significant, which is in increasing trend resulting from snow melt contribution. Snow melt contribution in stream flow increases approximately at rate of 2 per cent in winter, 5 per cent in summer and 4 per cent in annual flow and total runoff increases by 2 per cent in winter, 6 per cent in summer and 5 per cent in the annual per one degree centigrade temperature rise. The outputs of this study are important guidance for water resources managers to make and implement appropriate strategy for water resources management and hydropower development. It is also useful tool for adoptive planning *i.e.* more and efficient use of winter flow and mitigative and preventive measure for high flow periods. However the results presented here are not the predictions, but rather the model simulations on the snowmelt.

Acknowledgements

The authors are grateful to Deo Raj Gurung of ICIMOD for help in providing remote sensing data and processing data. Special thanks to Central Department of Environmental Science for providing research opportunity, and the others not mentioned here specifically, are gratefully acknowledged.

References

Adam, J.C., Hamlet, A.F., and Lettenmaier, D.P., 2009. Implications of global climate change for snowmelt hydrology in the twenty-first century. *Hydrological Processes*, 23: 962–972.

Akhtar, M., Ahmad, N., and Booij, M.J., 2008. The impact of climate change on the water resources of Hindukush-Karakorum-Himalaya region under different glacier coverage scenarios. *Journal of Hydrology*, 355: 148–163.

Braun, L.N., Grabs, W., and Rana, B., 1993. Application of a conceptual precipitationrunoff model in Langtang Khola Basin, Nepal Himalaya. *Snow and Glacier Hydrology*, 218: 221–237.

Bürgi, C. *et al.*, 2009. incorporating remote sensing data in modeling hydrology for the Himalayan catchment ETH Zurich, Institute of Environmental Engineering Hydrology and Water Resources Management Zurich, Switzerland.

Higuchi, K., Ageta, Y., Yasunari, T. and Inoue, J., 1982. Characteristic of precipitation during monsoon season in high mountain areas of the Nepal Himalaya (Proceedings of the Estern symposium, July 1992), IAHS pub. No. 138, pp. 21–30.

IPCC, 2007. *Climate Change 2007: Synthesis Report*. In: (Eds.) C.W. Team, R.K. Pachauri, and A. Reisinger. Geneva, Switzerland

Kayastha, R.B., Ageta, Y. and Fujita, K., 2005. Use of positive degree day methods for calculating snow and ice melting and discharge in glacierized basin in the Langtang Valley, central Nepal, climate and hydrology in mountaun areas, pp. 7–14.

Konz, M., Uhlenbrook, S., Barun, L., Shrestha A. and Demuth, S., 2005. Implementation of processes-based catchment model in a poorly gauged, highly glacierized Himalayan headwater. *Journal of Hydrology Earth System Science*, 11: 1323–1339.

Kundzewicz, Z.W., Mata, L.J., Arnell, N.W., Doll, P., Kabat, P., Jimenez, B., Miller, K.A., Oki, T., Sen, Z. and Shiklomanov, I.A., 2007. Freshwater resources and their management. In: *Climate Change 2007: Impacts, Adaptation and Vulnerability. Contribution of Working Group II to the Fourth Assessment Report of the Intergovernmental Panel on Climate Change*, (Eds.) M.L. Parry, O.F. Canziani, J.P. Palutikof, P.J.v.d. Linden, and C.E. Hanson. Cambridge University Press, Cambridge, UK, pp. 173–210.

Leavesley, G.H., 1994. Modeling the effects of climate change on water resources: A review. *Climatic Change*, 28: 159–177.

Nijssen, B., O'Donnell, G.M., Hamlet, A.F. and Lettenmaier, D.P., 2001. Hydrologic sensitivity of global rivers to climate change. *Climatic Change*, 50: 143–175.

Rango, A., 1997. The response of areal snow cover to climate change in a snowmeltrunoff model. *Annals of Glaciology*, 25: 232–236.

Seko, K., 1987. Seasonal variation of altitudional dependence of precipitation in Langtang Valley, Nepal Himalaya. *Bull. Glacier Res.*, 5: 41–47.

Shrestha, R., Takara, K., Tachikawa, Y., and Jha, R.N., 2004. Water resources assessment in a poorly gauged mountainous catchment using a geographical information system and remote sensing. *Hydrological Processes*, 18: 3061–3079.

Singh, P. and Bengtsson, L., 2005. Impact of warmer climate on melt and evaporation for the rainfed, snowfed and glacierfed basins in the Himalayan region. *Journal of Hydrology*, 300: 140–154.

Singh, P. and Kumar, N., 1997. Impact assessment of climate change on the hydrological response of a snow and glacier melt runoff dominated Himalayan river. *Journal of Hydrology*, 193: 316–350.

Singh, P., Haritashya, K.U.K. and Kumar, N., 2008. Modeling and estimation of different components of streamflow for Gangotri Glacier basin, Himalayas. *Hydeological Sciences Journal*, 53(2): 309–322.

Stewart, I.T., 2009. Changes in snowpack and snowmelt runoff for key mountain regions. *Hydrological Processes*, 23: 78–9410.

Wilson, W.T., 1941. An outline of the thermodynamics of snow-melt. *Trans AM Geophysics. Union*, Part 1: 182–195.

Chapter 5

Impact of Climate Change on Himalayan Wetlands: Adaptation and Management–A Case Study of Khajjiar Wetland, Chamba, Himachal Pradesh

☆ *Dinesh Kumar*

Recent scientific assessments indicate that, as the global temperatures continue to warm-up leading to climate change. The average global surface temperature has increased by 0.2°C per decade in the past 40 years, and global average precipitation increased 2 per cent in the last 100 years and it is further expected to increase in the range of 1.6°C to as much as 6°C by 2050. The climate at the end of this century will be substantially warmer than that of the past century. The joint effects of climate change will be either higher or lower precipitation, rise in temperature, and increased evapotranspiration.

Water, essentially, one of the most important elements for life, will be most heavily impacted by climate change. Longer spells of dry heat will cause water scarcity in the long run. Potential drought and various other vagaries are associated with climate change. Climate change will alter the amount of available water by magnifying the differences between the rainy and droughty periods, drastically decline the ground water recharge. It will lead to rapid drying out and loss of all sources of water. The water resources in any form and at any location will be affected. Climate change–water shortages–warmer temperatures are likely to increase. Spectrum of ecology, economy and society will be impacted by increasing temperatures, erratic rainfall, and various other vagaries of climate change. It is not surprising that the impact of this will be very fast. What is alarming, however, is that the consequences of this change are still unclear as well as the rate at which this process is advancing. Negative water balance will lead to falling water level in shallow lake. It might be induced by complex causes, but climate change is thought to be important contributor.

Due to climate change the unique biodiversity will come under tremendous pressure. As temperatures rise, it would lead to major changes for sensitive ecosystems, thus the living condition will be affected. The disturbances and threats can be broadly categorized as habitat destruction and degradation, loss of ecosystem integrity, depletion of species abundance and diversity. Even minor changes in temperature might have major impact on these sensitive ecosystems. Habitat loss may aggravate the situation which could upset the entire ecosystem as it will break the delicate balance of these sensitive ecosystems. Species' habitats will decrease, thus the chances for various ecosystems to adapt naturally will diminish. It might pose serious threats – not only to the continued maintenance and functioning of ecosystems and their biological diversity, but to its well-being.

Fresh water wetlands and swamps of the Himalayan region are also susceptible to the impacts of predicted climate change. Climate change–water shortages–warmer temperatures are likely to increase in near future. Spectrum of ecology, economy and society will be impacted by increasing temperatures, erratic rainfall, and various other vagaries of climate change. It will be difficult to adapt to the new condition. Warmer temperature as a result of climate change will affect the runoff and availability of water. It will have negative impact on water sensitive systems (ecology and economy). Sudden collapse of environmental and economic system may occur and it will aggravate the environmental crisis.

Today, wetlands as ecosystems stand on the fringes of our conservation policies and face the risk of being lost in the face of unplanned development and pressing resource needs. It is realized that in order to manage the wetland areas more efficiently and effectively, there must be the balance between ecological conservation and livelihoods of communities.

Wetland forms an important ecosystem essential for existence of life on earth. Wetlands, as vital ecosystems, have influenced the development of human societies since historical times. Floodplains of rivers and vicinities of wetlands are considered as the cradle of human civilization. Today, they provide staple food such as rice and fish for more than half of the human population. However we treat them with indifference rather than with care. As urbanization increased, the wetland also started vanishing. The people who had damaged the wetlands have already begun to pay the price in terms of scarcity of water, flooding, pollution, and toppling the micro climatic pattern.

Wetlands have long attracted the attention of public and scientists because of the charm, copiousness, visibility as well as for their recreational and economic importance. Wetland is a peculiar kind of landform called natural wonder. Each type of wetland has a unique function to perform *viz.*, kidneys of landscape, biological supermarkets, flood buffer, carbon sequestration etc. A shift from the view that 'wetlands are wastelands' and hence to be developed, to the growing awareness of its intrinsic values and functions have resulted in several international efforts to conserve, protect and restore wetlands in recent decades.

Wetlands are transitional areas between upland and deep water aquatic systems. By definition, wetlands are ecosystems whose physical, chemical and biological characteristics are determined by the constant or recurrent presence of water at or near the soil surface. The regulatory definition of wetlands is "land that has a predominance of hydric soils, and that is inundated or saturated by surface or groundwater, a prevalence of hydrophytes' vegetation typically adapted for life in saturated soil conditions"

The term wetland ..refers to lowlands covered with shallow and sometimes temporary or intermittent waters. Wetlands are "lands transitional between terrestrial and aquatic systems where the water table is usually at or near the surface of the land or the land is covered by shallow water".

According to the *Ramsar definition,* wetlands include a wide range of habitats and ecosystems. The term "wetlands" groups together a wide range of inland, coastal and marine habitats which share a number of common features.

Wetlands come in all shapes and sizes and go by names such as marshes, swamps, scrub-shrubs, bottomlands, oxbows, or sloughs.

All wetlands have three things in common:

1. A soil that is at least periodically saturated or pounded and exhibits certain water-loving plants is used as one indicator to identify wetlands. Two other indicators must be present to identify wetlands: hydric soil and hydrology.
2. Water to create pounding or saturated conditions of the soil in the upper layer during the growing season.
3. Vegetation that can tolerate anaerobic conditions;

According to *Ramsar convention* wetlands are areas of marsh, fen, peat land or water bodies, whether natural or artificial, permanent or temporary, static or flowing, fresh, brackish or salt including areas of marine water, the depth of which at low tide does not exceed six meters.

There are various types of wetlands and many development options depend on the appropriate choice of wetland types. Therefore, it is important to have sufficient knowledge about the types, extent and distribution of wetlands, production levels and the potential functions and values of wetlands and wetland resources.

Objectives

The main objective of the present chapter is to highlight the possible impact of changed climatic conditions on wetland adaptation and management strategies. It seems that no research linking the policy impacts to biophysical and then socio-economic change has been carried out. Therefore, the overall picture of impacts of policies on biophysical and socio-economic settings is not presented for policymakers so that they can optimize the policies to maximize social welfare. That leaves a gap in knowledge of total economic values of wetlands.

The objectives of the chapter are to highlight the following aspects in short:

- ☆ Evaluates the impact of climate change, more particularly its geo-ecological implications.
- ☆ Examine the wetland of the Himalayan region to the impacts of climate change.
- ☆ Suggest the method of adaptation to climate change and assist in formulating an integrated strategy for the adaptation to climate change.
- ☆ Prepare guidelines for 'regional adaptation programs.'
- ☆ Delineate various geo-morphological features. Identification of different habitat, wetland monitoring and mapping;
- ☆ It also proposes to study the land use patterns–to define the effects of human utilization of the ecosystem and the utilization of the water bodies by local inhabitants.
- ☆ Provide management recommendations for the conservation of the wetland.

Approach of Study

A case study approach is used to examine the adaptation and management of the Himalayan region to the impacts of climate change. In this research, the main question is how the negative impact of climate change is going to affect the wetland. As the ultimate objective of the study is to suggest the method of adaptation to climate change, a case study approach is appropriate in formulating 'regional adaptation programmes.'

Study Area

The Khajjiar Lake in the district has recently been listed among 94 wetlands of India covering 23 states and one Union Territory which have been identified under the National Wetland Conservation Programme (NWCP) for carrying out conservation activities. The Khajjiar wetland is one of the tectonic wetlands like Nainital and Bhimtal in Uttarakhand. Khajjiar Lake and its surroundings are one of the most picturesque saucer shaped plateaus. The watershed is located at a height of 1,940 m in a valley between Dhauladhar and Pir Panjal ranges of the Himalayas. The watershed lies at latitude of 32.5° N and longitude of 76.1° E. The size of the lake is not large and is more or less like a pond of about 60 to 80m radius. The green pastures surrounding the lake are approximately 1.5 km long and 1.0 km wide. The lake region experiences a moist temperature climate. Precipitation is in the form of snow during January to March and rain during the south-west monsoon. The annual average precipitation is 1.2 m. The soils are sandy loam in the meadow part, while under the forest they are highly organic with humus and litter. The soil thickness varies from 30 to 40 cm. Khajjiar is, home to the elusive serow and the common Black Bear and numerous well laid out trekking paths is a star attraction for the wildlife tourist.

Socio-Cultural Aspects of Khajjiar Wetland

Khajjiar is famous for the popular Khajji Naga shrine dedicated to the serpent god from which the name is believed to have been derived. The temple dates back to the 10th century and is interspaced with different patterns and images on the ceiling and wooden posts. A curious blend of Hindu and Mughal styles of architecture is reflected in the wooden carvings on the ceilings and wooden posts. The image carvings are said to represent the Kauravas who were tied up here in the hideout by the Pandavas. The temple consists of a spacious congregation hall sufficiently enclosed by wooden supports. The dome-shaped shrine is made of slates locally extracted from limestone quarries. Adjoining are other shrines of Shiva and Hadimba goddess also. The age-old tradition of sacrificing a goat is still practiced outside the temple premises. The lake is considered to be holy by the locals as legend associated it with Khajinag, the deity in the Khaaji temple. The lake has been neglected over the year by year.

Khajjiar Wetland

Geo-environmental Factors

The lake is gradually silting, partly because of reduced inflow, but largely through human modification of the catchment area and local developments in Khajjiar.

Geographical Condition

The watershed is located at a height of 1,940 m in a valley

between Dhauladhar and Pir Panjal ranges of the Himalayas. The watershed lies at latitude of 32.5° N and longitude of 76.1° E. The size of the lake is not large and is more or less like a pond of about 60 to 80 m radius. The green pastures surrounding the lake are approximately 1.5 km long and 1.0 km wide.

Climate, Soil, Vegetation

The lake region experiences a moist temperature climate. Precipitation is in the form of snow during January to March and rain during the south-west monsoon. The annual average precipitation is 1.2 m. The soils are sandy loam in the meadow part, while under the forest they are highly organic with humus and litter. The soil thickness varies from 30 to 40 cm.

Habitat Diversity

Khajjiar is home to the elusive serow and the common Black Bear.

Function

Wetlands are efficient water filters, because wetland plants and soils clean water of contaminants before it moves into surface waters or groundwater. Wetlands filter runoff and remove sediment, nutrients, pesticides, metals, and other pollutants.

Threats

The level of pollution is now having a drastic effect on the number and variety of waterfowl utilizing the lake, and must be having a serious impact on fish production.

Conservation Prospects

(a) The team recommended treatment of catchment area by taking vegetative and mechanical measures, management of excreta of the livestock near the water body and awareness about wetlands' multifunctional role. (b) the creation of permanent shallow-water areas where reed-beds and other perennial aquatic vegetation will develop and attract additional bird species; (c) the construction of floating islands which can be tethered in deep water and provide roosting and perhaps nesting sites for other species.

Wetland Values

Wetlands provide many services and commodities to humanity. Regional wetlands are integral parts of larger landscapes, their functions and values to the people in these landscapes; depend on both their extent and their location. Each wetland thus is ecologically unique. Wetlands perform numerous valuable functions such as recycle nutrients, purify water, attenuate floods, maintain stream flow, recharge ground water, and also serve in providing drinking water, fish, fodder, fuel, wildlife habitat, control rate of runoff in urban area, buffer shorelines against erosion and recreation to the society. The interaction of man with wetlands during the last few decades has been of concern largely due to the rapid population growth–accompanied by intensified industrial, commercial and residential development further leading to pollution of wetlands by domestic, industrial sewage, and agricultural run-offs as fertilizers, insecticides and feedlot wastes. The fact that wetland values are overlooked has resulted in threat to the source of these benefits. Wetlands are often described as "kidneys of the landscape"

Functions of Wetlands

Wetlands are among the most productive ecosystems. They directly or indirectly support millions of people and provide goods and services to them. Various goods and services provided by wetlands are as follows:

- ☆ Support all life forms through extensive food webs.
- ☆ Habitat to aquatic flora and fauna, as well as numerous species of birds, including migratory species.
- ☆ Filtration of sediments and nutrients from surface water.
- ☆ Nutrients recycling.
- ☆ Water purification.
- ☆ Floods mitigation.
- ☆ Maintenance of stream flow.
- ☆ Ground water recharging.
- ☆ Provide drinking water, fish, fodder, fuel, etc.
- ☆ Control rate of runoff in urban areas.
- ☆ Buffer shorelines against erosion.
- ☆ Comprise an important resource for sustainable tourism, recreation and cultural heritage.
- ☆ Stabilization of local climate.
- ☆ Source of livelihood to local people.

- ☆ Genetic reservoir for various species of plants (especially rice).

Functions and Values of Wetlands

Water Quality

Wetlands are efficient water filters, because wetland plants and soils clean water of contaminants before it moves into surface waters or groundwater. Wetlands filter runoff and remove sediment, nutrients, pesticides, metals, and other pollutants.

Erosion Control

Wetlands reduce the severity of floods by acting as natural retention areas; Consequently, destruction or loss of wetlands makes downstream flooding worse.

Nurseries for Fish

Wetlands serve as important nurseries for numerous fish, as many species depend on wetlands for some or all of their life cycle, where they feed, take refuge, or reproduce.

Wildlife Habitat

Wetlands are home to many different animals and, importantly, are stopover or winter habitat for many migrating birds.

Recreation

Many recreational activities depend on wetlands, including fishing, hunting, bird-watching, boating, and wildlife photography.

Cultural Importance

Many cultures have been heavily influenced by or rely upon wetlands, wetlands rely upon those wetlands for fishing, hunting, and trapping. Similarly, wetlands are important to several indigenous tribes.

Management Objectives

The Management needed to mitigate the threats and sustain the assets for these significant wetland areas and are summarized below:

- ☆ Protect and enhance native vegetation.
- ☆ Protect and enhance socio-economic values.
- ☆ Protect and enhance native fauna and their habitat.

Physical, Biological and Management-Related Threats to the Values of Wetlands

Physical Threats

Threats	*Impact*	*Response to Reduce Threat*
Increased erosion in the catchment	– Loss of topsoil or wetland soil. – Reduced productivity of the soil. – Decreased water quality (sedimentation, nutrients)	– Identify areas of erosion within wetland area – Increase buffer zones – Investigate erosion and sediment transport mitigation techniques for identified areas *e.g.*
Modified Hydrology -Drainage	– Changes base water level of the wetland – Causes upstream erosion – Affects plant composition – Changes hydrology	– Determine volume of water extraction and its impact on the wetland. – Determine volume of water extraction and its impact on the wetland
Increased acidity (soil)	– Soils light and cleared – Sandy with low fertility and acidity – Poorly managed	– Investigate whether acidity is a major issue effecting wetland soils – Investigate methods to mitigate soil acidity (if necessary)
Impacts associated with climate change (temperature, evaporation and rainfall)	– Increasing temperatures – Changes in rainfall – Higher evaporation rates – Possible increase in bushfires – Alters seasonal weather patterns – Less available water – Impact on native vegetation – Impact on local communities living in rural areas	– Investigate the Water balance for the wetland – Determine volume of water extraction and its impact on the wetland – Investigate number of farm dams in catchment and holding volume – Reduce carbon emissions? – Water efficiency

Biological Threats

Threats	*Impact*	*Response to Reduce Threat*
Poor water quality	– Increase in nutrients and pesticides from surrounding land use decreases water quality – Change in wetland plant communities – Sedimentation of the wetland	– Establish wetland buffer zone to filter water running off from catchment into the wetland
High Abundance of native fauna	– Grazing impact on native plants – Adding stress to native plants (in combination with grazing cows/rabbits etc)	– Increase habitat areas – Remove feral species to reduce competition – Education of habitat requirement.
High abundance of feral fauna	– Reduced habitat for native fauna. – Planting of introduced plants compete with natives – Decreased existing vegetation	– Investigate viability of current seed, for regeneration – Map vegetation and weeds to help decide priority actions – Identify appropriate species list for planting – Identify areas for regeneration.

Management

Threats	*Impact*	*Response to Reduce Threat*
Non-sustainable grazing	– Trampling and compaction – Reduced biomass (groundcover) allowing for open space and weed invasion – Stock brings in weed seeds from surrounding land. – Overstocking and overgrazing exacerbate the impacts	– Identify areas of undesirable grazing – Fencing areas from grazing – Provide off-stream watering points – Identify appropriate grazing techniques *i.e.* rotational grazing

Management

Threats	*Impact*	*Response to Reduce Threat*
Impacts associated with sand mining	– Erosion – Modification to soil properties – Exposing sub soils – Runoff – Sedimentation – Groundwater	– Investigate impact on Paddy's River swamps – Increase buffer zones (if necessary) – Investigate Erosion mitigation techniques (if necessary)
Forestry	– Clearing of surrounding land (every 30 years when pine is harvested) – Creating roadways – Increased run-off and sediment input from roadwaysand clearing – Limited riparian vegetation – Increased access to wetland areas for recreation	– Increased weed invasion *i.e.* pine encroachment – Input soft engineering techniques (already occurring) – Create rollovers to divert run-off (already occurring) – Maintain wildlife corridors for animal migration-buffers around wetland reduce run-off filtration and to act as a fire buffer
Land Development (subdivision, golf courseroad)	– Disturbs surrounding soils – Increased pollutants and wastes – Higher sediment and nutrient loads entering wetland – Increases spread of invasive species (native and introduced) – Increases run-off – Increased pressure of use on the wetlands – Possible increase of water extraction from wetland Changes surrounding vegetation	– Carefully evaluate development applications – Evaluate land zoning and lot sizes – Ensure all pollution mitigation techniques implemented – Create reserves, with areas of restricted access – Increase buffer zones – Educate local community about their impacts on the surrounding wetlands

Management

Threats	*Impact*	*Response to Reduce Threat*
Threat to Aboriginal cultural heritage	– Desecration of artifacts – Desecration of a significant site	– Public education to minimize impact – Restrict access to significant area
Recreation	– Increased number of visitors – Boats causing wave erosion	– Public education to minimize impact – Restrict access to significant areas
Uncoordinated approach to management	– Government vs. individual landholder management- Responsibility of individual – Uncoordinated approach to management threats	– Identify roles and responsibilities of landholders – Develop co-ordinate approach to managing threats across all land within the wetland area Develop communications strategy for clear communication between land managers

- ☆ Understand the impact of the catchment processes on water quality.
- ☆ Support sustainable land use.
- ☆ Increase community awareness of threats and values of Wetlands
- ☆ Implement a collaborative approach to wetland management

Protect and Enhance Socio-Economic Values

- ☆ Maintain sites significance.
- ☆ Maintain recreation use.
- ☆ Maintain community access.
- ☆ Identify and implement sustainable land use.
- ☆ Identify and protect significant areas.
- ☆ Identify recreation use and options for maintaining and minimizing its affects.
- ☆ Ensure access remains open to the public where appropriate.
- ☆ Maintain economic viability of surrounding land.
- ☆ Increase awareness of the importance of the wetland and the surrounding area.

Major Threats to Khajjiar Wetland

- ☆ Siltation: Deposition of silt eroded from far off hill catchments and immediate areas has reduced the lake area.
- ☆ The storage capacity of this lake has considerably reduced.
- ☆ Encroachments: Wetland has been encroached upon by unauthorized in surrounding area.
- ☆ Public Pollution from shops and hotels activities
- ☆ Fishing and Poaching: However illegal fishing and poaching does take place and it has been a cause of concern.
- ☆ Grazing: Indiscriminate grazing in the catchments and right in the wetland zones is damaging the wetland ecology.

Wetland Management: Current Status

Wetlands are not delineated under any specific administrative jurisdiction. The primary responsibility for the management of these ecosystems is in the hands of the Ministry of Environment and Forests. Although some wetlands are protected after the formulation of the Wildlife Protection Act, the others are in grave danger of extinction. Effective coordination between the different ministries, energy, industry, fisheries revenue, agriculture, transport and water resources, is essential for the protection of these ecosystems.

Protection Laws and Government Initiatives

Wetlands conservation in India is indirectly influenced by an array of policy and legislative measures. Some of the key legislations are given below:

- The Indian Fisheries Act – 1857
- The Indian Forest Act – 1927
- Wildlife (Protection) Act – 1972
- Water (Prevention and Control of Pollution) Act – 1974
- Territorial Water, Continental Shelf, Exclusive Economic Zone and other Marine Zones Act – 1976
- Water (Prevention and Control of Pollution) Act – 1977
- Maritime Zone of India (Regulation and fishing by foreign vessels) Act – 1980
- Forest (Conservation act) – 1980
- Environmental (Protection) Act – 1986
- Coastal Zone Regulation Notification – 1991
- Wildlife (Protection) Amendment Act – 1991
- National Conservation Strategy and Policy.
- Statement on Environment and Development – 1992
- National Policy and Macro level Action Strategy on Biodiversity-1999.

Management of Common Property Resources

Public administration plays a vital role in the sustainability of common property resources. Efficiency and economy in the use of common property resources are the components of a conceived plan for a sustainable living. The government owes the responsibility of managing wetlands. It administers a range of social, economic and environmental programmes that impact on conservation and the wise use of resources. But government fails to take the lead role in wetland conservation as best performed in cooperation with the community. There is an urgent need of a public policy aiming at wetland conservation within the broader framework of environmental management.

A successful wetland policy is one that recognizes the need to harmonize with various other policy instruments, goals, objectives and strategies. The wetland governance could succeed only when it is committed to working cooperatively with all spheres of the government to achieve sound wetland management outcomes. Considerable benefits can be gained from governments (Central, State and Local) working together to share approaches, solutions and resources for achieving common outcomes.

References

Agrawal, H.O. and Thakur, R.S., 1987. Analysis of some lake ecosystems of Himachal Pradesh with special reference to their conservation and management. Final Technical Report MAB Research Project No. 20/32/81 135pp. Submitted to Ministry of Environment and Forest, Govt. of India, New Delhi.

Alford, R.A., Bradfield, K.S. and Richards, S.J., 2007. *Global Warming and Amphibian Losses.*

Directory of Indian Wetlands, 1993. WWF–India Asian Wetland Bureau.

Dayanand, G.Y., Hosetti, B.B. and Sulubha, P., 2001. Water quality evaluation for Gudavi bird sanctuary. In: *Trends in Wildlife Biodiversity Conservation and Management*, (Eds.) B.B. Hosetti. Daya Publishing House, New Delhi.

Gandhi, S.S. and Singh, S.K., 1995. Avifauna of Asan Barrage. *Cheetal*, 34(1): 29.

Joshi, P.C. and Joshi, N., 2004. In: *Biodiversity and Conservation*. APH Publishing Corporation, New Delhi.

Kumar, Arun, 1998. Habitat evaluation of migratory waterfowls using remote sensing technique. Project Report. Zoological Survey of India, Dehradun.

Narang, M.L., 1994. In Birds (Aves). *Fauna of Conservation Area,* 5: Rajaji National Park, Zoological. Survey of India, 25–53.23.

Narang, M.L., 1990. Migratory water birds of Dhalipur lake near Dehradun (U.P.). *Indian Journal of Forestry,* 13(1): 63.

Pandey, S., 1993. Changes in waterbird diversity due to the construction of Pong Dam reservoir, Himachal Pradesh, India. *Biological Conservation,* 66(2): 125.

Singh, A.P., 2000. Birds of Lower Himalayas: Dehradun Valley and Neighboring Hills. *Forktail,* 16: 101–123.

Singh, S., Kothari, A. and Pande, P. (Eds.), 1990. *Directory of National Parks and Sanctuaries in Himachal Pradesh.* Indian Institute of Public Administration, New Delhi, pp. 67–69.

Pandey, S., 1993. Changes in Waterbird Diversity due to the Construction of Pong Dam Reservoir, Himachal Pradesh, India. *Biological Conservation,* 66(2): 125.

Wake, D.B., 2007. Climate Change Implicated in Amphibian and Lizard Declines. *Proc. Natl. Acad. Sci., USA,* 104: 8201–8202.

Chapter-6

Scramble for Water in Himalayas: A Case Study of Water Looting in Ravi Basin in Himachal Himalayas

☆ *Mohinder K. Slariya*

ABSTRACT

The potential capacity of Ravi basin in Chamba district is 2301.5 MW, out of which 1038.5 MW potential has tapped through four commissioned power projects. There are 50 mini and micro hydro-electric power projects sanctioned for Ravi basin. Out of these there are 7 mini power projects have been sanctioned in 22 kms long Saal River. Whole Saal valley has received/is receiving bad impacts on the lives/livelihood of thousands of villagers as well as on the physical environment. The area presently has stricken by the new emerged phenomena of cloud burst which is the outcome of the disturbance created by these power projects. The disturbances in the lives of thousands of people and fragile physical environment of the area can be saved by using the Dr. Anil Joshi's model of converting water mills into electricity producing units. An attempt has been made by the researcher to present an alternative model for the Saal valley hydro-electric developments after presenting factual position of the area.

Keywords: *Ravi basin, Small/mini/micro power projects, Water mills, Biodiversity.*

Man is the finest creation of God and he has basic right to live in that environment in which he can lead a peaceful life with discipline, with a sense of duty to preserve and conserve the environment for the future generations by making effective planning in respect of air, water, land, hills, trees, plants and wild animals especially natural eco-system. Since time memorial he has totally been dependent on *"Mother Nature"* for his basic needs. Since ages, man has involved himself in making his life full of comforts. In search of better living style with the development of civilization, day back in 1672 Bon Guericke of Germany has invented electricity. Right from that time to date the electricity has got the status of essential commodity of our routine life.

The potential capacity of hydropower generation in India is 1,84,000 MW out of which only 15 per cent has been tapped and 7 per cent is on the way of its commissioning; totaling thereby 22 per cent has been either commissioned or commissioning. In Himachal Pradesh, there are five natural drain systems in form of five river basins having 25, 000 MW potential capacity of hydro power generation (CM Budget speech of 2005-06). The potential capacity of Ravi River and its tributaries in Chamba district is 2301.5 MW *(R.L. Singh, 2004)* out of which 1038.5 MW potential has tapped through four commissioned power projects *i.e.* Bhuri Singh Hydro-electric Power Project of 0.5 MW, Baira Suil Hydro-electric Power Project of (198 MW), Chamera-I Hydro-electric Power Project of (540 MW) and Chamera-II Hydro-electric Power Project of (300MW).

Hydro-electric power developmental activities are in progress in the Peer Panjal ranges of young Himalayas, which are in formation stage, for the last thirty years. The whole area of Chamba district falls in seismic zone V and in this zone disturbance in physical environment is not desirable particular the tunnelization of natural routes of the rivers and the researvoirization of the water on the rocks, which has no experience of bearing weight. The rocks may weaken which may have very serious consequences. The affected people have not been rehabilitated they have to go in the court of law after sacrificing everything, once which was their own. In such type of developments environmental, social, humanitarian and ecological aspects has always been ignored.

Table 6.1: Showing Disturbance Created by Installation of some Power Projects in Himachal Pradesh

Name of the Power Project	*Name of the River*	*Installation Capacity (in MW)*	*Head Race Tunnel (in KM)*	*Tail Race Tunnel (in KM)*	*Total Diversion (in KM)*
Nathpa-Jhakri	Satluj	1500	27.394	0. 982	28.376
Rampur	Satluj	0415	15.088	–	15.088
Baspa-II	Baspa (satluj)	0300	07.095	0.475	07.570
Luhri	Satluj	0775	38.138	0.454	38.592
Dehar	Beas	0990	06.065	–	06.065
Parvati-II	Parvati (Beas)	0800	31.052	0.240	31.292
Parvati-III	Parvati (Beas)	0520	07.098	2.007	09.105
Chamera-I	Ravi	0540	06.400	2.004	08.404
Chamera-II	Ravi	0300	07.083	3.046	10.129
Chamera-III	Ravi	0231	15.093	–	15.093
Baira-suil	Baira-suil (Ravi)	0193	07.063	–	07.063
	Total	**6564**	**167.569**	**9.208**	**176.777**

Source: Author's compilation from different power projects in Himachal Pradesh.

After exploiting Ravi 1038.5 MW, now State Government has made target the tributaries of Ravi (small nallas and rivulets). In Chamba district, there are 15 mini, small and micro power projects in different stage of their execution in the upstream of Chamera-I (R.L.Singh, 2004). According to Himurja there are 50 mini and micro hydro-electric power projects. State Government has signed MoU with different executing agencies and is in different stages of their execution (www.himurja.nic.in).

Table 6.2

Name of the Power Project and Year of Commissioning	*Number of Displaced Families*	*Submerged Area (in Hectare)*	*Lengh of the Reservoir (in Kms)*
Pong dam (1962)	16,352	045,000	42
Bhakra (1963)	36,000	168,000	84
Chamera-I (1994)	01,554	039,000	29
Chamera-II (2003)	00,200	–	02.6
Total	**54,106**	**2,52,000**	**157.6**

Nearby Chamba the historic Saal River, one of the major tributary of Ravi, flows in rain and snow fed hydrological homogenous region, area roughly defined between the elevations 1500 meter to 3000 meter. It is being exploited to the greater extend by disturbing the lives of the natives, environment, playing with the sentiments of the people because of the fact that confluence point of Ravi and Saal river is known as *Ganga* for the poor people of the area, occupation (the area is known for growing seasonal as well as non- seasonal vegetables, which has checked the out village migration of the local youth); above all the fragile eco- system including flora and fauna of the area. The total length of the Saal River is 22 kilometers. On this river first power project of Chamba and second power project of Asia, after Darjling is functioning in the name of Bhuri Singh mini

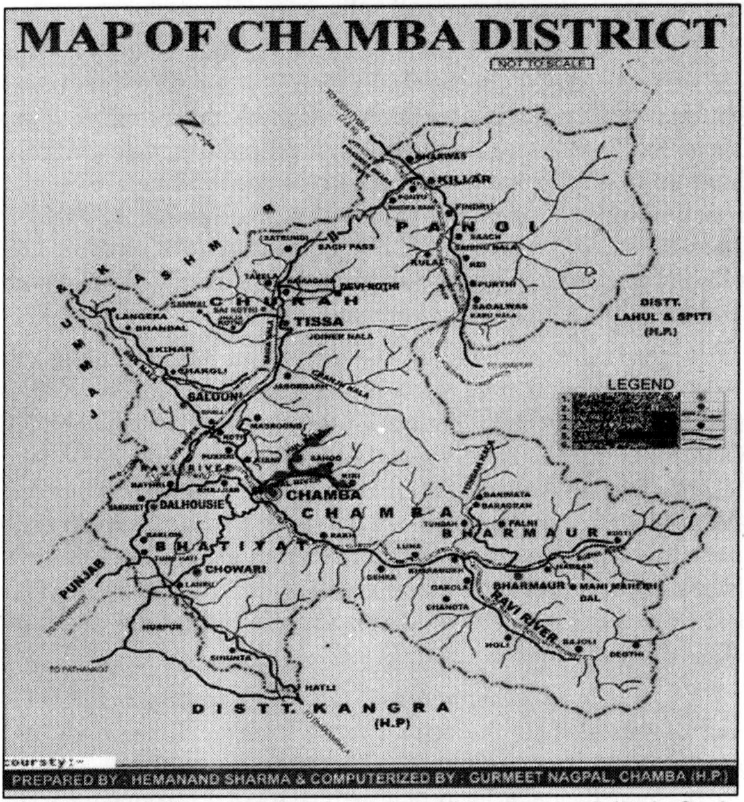

***Red line Saal River and dots indicates mini power projects in Saal River, major tributary of Ravi.**

hydro-electric power project. Saho hydel electric power project of 5 MW is on the way of its completion, along with this, DPR of Hul hydel electric power project of 4.5 MW capacity has been submitted and environmental clearance has been sought at present, it is ready for installation, Saal HEP stage-II of 2.5 MW, Saho HEP stage –II of 3 MW, Hul HEP stage-II of 1.40 MW, Kiri HEP of 1 MW are indifferent stages implementation (www.himurja.nic.in).

On the one hand whole valley is going to develop from heritage tourism point of view and on the other hand, the natural beauty of this valley is going to be disturbed by these serials of power projects in the adjoining river rivulets of Chamba. The natural beauty of the Ravi has already been sacrificed by Chamera-I and Chamera-II HEP projects, particularly during summer season. In this season Ravi has been converted into a small nalla.

There is another untouched and unexploited source of hydro electricity generation in the Saal River. The whole is agriculture dominated having self-dependency and the carrier of *"Hatt and Gharat"* culture. According to a rough estimate made by HESCO there are nearly 50,000 water mills in the entire Himalayas region from the Northeastern states to J&K, which can produce 2500 MW, assuming each generates 5 KW, that is a cash generation of Rs. 1200 million per hour and gives direct employment to 1,50,000 people assuming three persons per water mill.

According to an estimate made by gram panchayat and Zila Parishad member that there are 49 functional water mills in Zadera nalla on which Hul HEP project is proposed to be installed and 35 water mills in Saho nalla which became Saal at Chaminu village. According to Dr. Anil Joshi of HESCO a water mill can produce 5-10 KW power after slight modifications in the traditional water mills with little specifications with in the small expenditure of 35-45 thousand per water mill. There are 84 water mills in the Basin of Saal River, which can be modernized and set an example for the whole district as well as for the state.

The whole Saal region can be develop as model from tourist point of view along with electricity generation and flour grinding by providing direct employment to a number of local youth and will also preserve the pahari way of life which revolves around the water mills. Above all, all this can happen without disturbing the environment, ecosystem and over all the lives of people of the area are not disturb. Water mill cover and protected the cluster of trees which strengthen the soil and the problem of soil erosion of the Saal river can be solve and furthermore it will excel the natural beauty of the area which attracts the tourists along with electricity generation.

There are 84 water mills in the area, which can produce 420 KW power assuming 5 KW per water mill. 420 units electricity can be generated per hour through which Rs. 420 x 2.40=1008=00 per hour can be earn, this amount goes to Rs. 24,192=00 per day and Rs. 72,576=00 per month and Rs. 8,70,912=00 per year additional along with grinding. It will give direct employment to at least three persons per water mill (84x3=252 persons) and indirectly it increases manifolds.

This all can be done without disturbing anything. The conflict between development and the native people can be resolved easily. There will be no disturbance in the physical environment of the seismically sensitive zone-V, by adopting such type of approaches we can strengthen the physical environment and the possibility of the earthquakes in the region can be ruled out to a greater extent. The possibility of unrestness among the youth of the area can be ruled out easily by not snatching their traditional occupation of agriculture, which largely depends on the local kuhul (canal). The small patches of local varieties of trees around the water mills will increases the natural beauty of the area and invites tourists as well as researchers from far of places.

Saal River flows in snow and rain fed hydrological homogenous region and mostly depends on rain and snow for water. Because of drastic changes in climatic conditions of Ravi catchments areas whole area has experienced very less snow and rainfall during last ten years. In rainy season, Whole River experienced flood and damaged all flora of its catchments. There is no permanent or big glacier, but only seasonal snow, which are decreasing day by day. Blasting, five –five kilometer long open water channels and tunnels are not environmental friendly, particularly, in seismic zone-V such activities are not desirable and viable. The excess exploitation of this historical river may dry this river. The beautiful 22 kilometer long Saal valley may convert into dry valley; the scenic beauty may not be no more if the present situation persists. The lives of thousands people, residing in this valley may be disturbed and even the lives of people of Chamba town may be questioned by not fulfilling the need of the potable water. The wild animals may entered in the town because of blasting and extinction of flora of the area, as happened with Shakti Singh a resident of Surara mohalla of Chamba town of dated May, 6, 2006 (as told by him to the researcher). These innocent wild animals may be shooted down by the civilized people, because the civilized people in the name of development have snatched the natural habitat of the wild animals. It is high time to think over such developmental policies, in which man, nature and natural creation of the God may live in harmonious environment.

Today, we need to think again on the policies, which have been made for the hills because all the existing policies for the development of the hills are full of conflicts among the development and native people. We should think and make policies by taking local circumstances into consideration, so that this conflict can be resolved and the true fruits of development can be enjoyed. The feeling of insecurity among the local people of being disturbed by any developmental activities must not be there. If they disturbed by any developmental activity they must resettle in prior conditions, conducive for living, before starting and developmental project. In the policies minimum disturbance to eco- system as well as to social system must be insured. For all that what we need, a strong political will and the will power of the local people, because it is the political will that may brought any type of change in society and the political will can be implemented only through strong will power of the local people.

Chapter 7

Remedial Measure of Tapi River Flood at Surat City: A Study

☆ *Krupesh Chauhan, Jagruti Shah, Amol Pingale and Gaurang Desai*

ABSTRACT

Recent events such as hurricanes, tsunamis, earthquakes, power outages, floods and the threat of pandemics have highlighted our vulnerability to natural disasters. This vulnerability is exacerbated by many organizations' increasing dependence on computer, telecommunications, and other technologies, and trends toward integrating suppliers and business partners into everyday business operations. In response to these, many organizations are implementing disaster recovery planning processes.

In this paper we discuss how to identify threats and scenarios; how to articulate the disaster recovery strategies; and four elements of the generic disaster recovery plan: mitigation, preparedness, response, and recovery, with respect to Surat floods 2006. We try to provide examples of software that can help disaster recovery professionals in the planning and implementation process. Finally we present some trends that will reinforce the criticality of the issue.

Keywords: *Disaster recovery planning, Business continuity planning, Risk assessment.*

Introduction

Surat city is situated at the delta region of river Tapi (India). The river Tapi is originating from a mountain Satpuda and flowing through three states Maharastra, Madhya Pradesh and Gujarat. The river Tapi is merge in the Arabian Sea at about 15 kms away from Surat city. The map of Tapi river basin is shown in Figure 7.1. Ukai dam controls the flow of water and water level in the river Tapi, which is 100 kms away from Surat city. The dam is constructed at Ukai, Tal. Songadh in Distt. Surat. It is constructed for irrigation purpose mainly and also served the purpose of flood control, generation of hydropower and supply of industrial and drinking water. The average rainfall in the catchment area is about 785 mm and average yearly run off is 17,226 MCM. The salient features of Ukai Reservoir are shown in Table 7.1. Total seventeen rain gauge stations are available at different places and twenty two wireless stations are linked with the dam site, which gives the information about rainfall to the Central Water Commission. The Indian

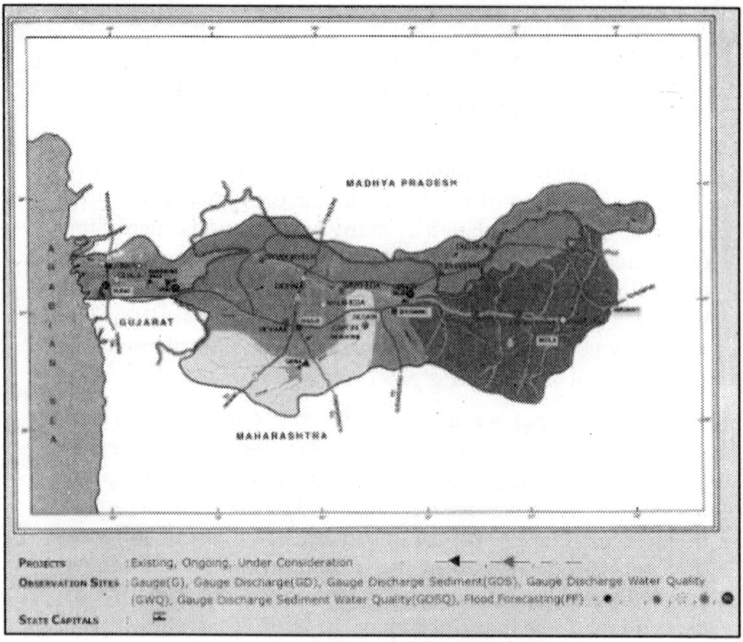

Figure 7.1: Map of Tapi River Basin

Meteorology department predicts the probability of rainfall in the catchment area during next 24 to 48 hours and same information is transferred to Central Water Commission and media. Flood occurs at Surat city frequently due to sudden release of water from Ukai dam in river Tapi. At the time of floods in river Tapi, Surat city and surrounding region are most affected. History of floods at Surat city, details of floods and remedial measures for prevention/minimization of the effect of floods at Surat city are discussed in the subsequent topics. A flood reduction plan is also prepared and presented.

Table 7.1: The Salient Features of Ukai Reservoir

Sl.No.	*Item*	*Description*
1.	Location	Songadh (District Surat)
2.	River	Tapi
3.	Catchment area	62225 sq.km
4.	Design flood discharge	49490 Cumecs
5.	Year of completion	1972
6.	Gross storage capacity	8.511 BCM
7.	Live storage capacity	7.092 BCM
8.	FRL	105.22 mt (345 ft)
9.	HFL	107 mt (351 ft)
10.	MDDL	82.3 mt (270 ft)
11.	River bed level	47.87 mt (157.05 ft)
12.	Install power	305 MW
13.	Hydraulic head	34-57 mt
14.	Maximum discharge	213 cumecs

Floods at Surat City

The Surat city has faced many floods since 1883. The history of the floods is given in Table 7.2.

Scenario of Flood during August 2006

The flood occurred in the year 2006 was devastating. The level of water started rising in the river tapi from 1st August 2006 and started spreading in the nearby area of city. By 5th to 9th August 2006, almost 90 per cent area of the city was flooded and the depth of flood water observed in the different areas was varying according to the topography of the City. The map of Surat city (Figure 7.4) is showing the variation in the depth of submergence in different colours. The information and warning about flood must be reached to the people timely so that the people can take their own measures of safety and precautions. Figure 7.2 shows the status of warning in Surat during flood 2006. From the figure, it is evident that about 43 per cent of people did not receive any warning from the Surat Municipal Corporation (SMC) or any other state agency. They learnt about the approaching flood only when they saw the water rising. This proportion was highest (64 per cent) in areas like citylight, umra and piplod, all of which are considered Surat's posh or upcoming localities. Only around 7 per cent of respondents said that they had received some warning from the administration through vans or Short Service Messages (SMS) on mobile phones. As many as 30.7 per cent of respondents said that they first learnt

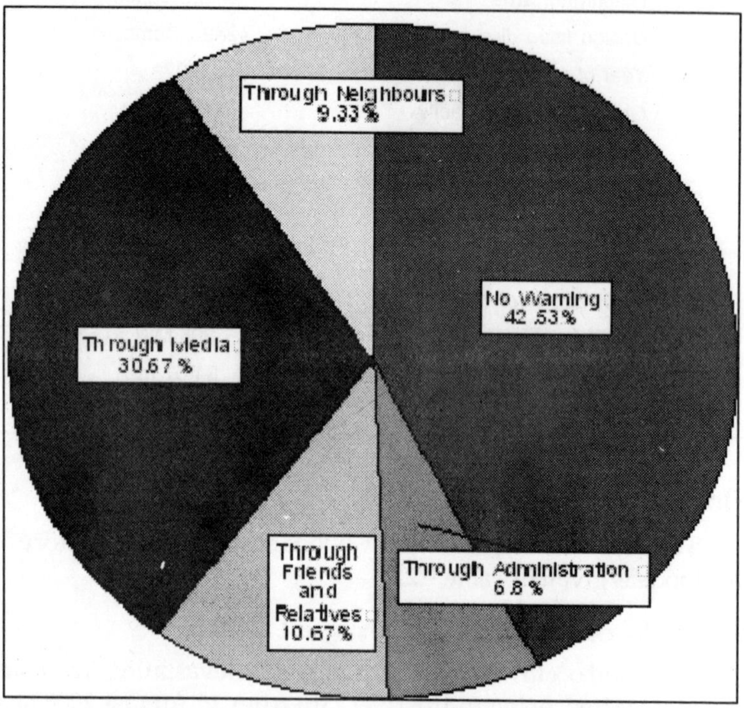

Figure 7.2: Pie Chart Showing Status of Warning (Source: Patel, 2002)

Table 7.2: Flood History at Surat

Sl.No.	Flood Event	Discharge (Lacs cusecs)	Water Level at Hop Bridge (mt)	Period
1.	1883	10.05	11.05	July
2.	1884	8.46	10.05	September
3.	1894	8.05	10.33	July
4.	1942	8.60	10.56	August
5.	1944	11.84	11.32	August
6.	1945	10.24	11.09	August
7.	1949	8.42	10.49	September
8.	1959	12.94	11.55	September
9.	1968	15.5	12.08	August
10.	1994	5.25	10.10	Aug.- Sep
11.	1998	7	11.40	September
12.	2006	9.09	12.40	August

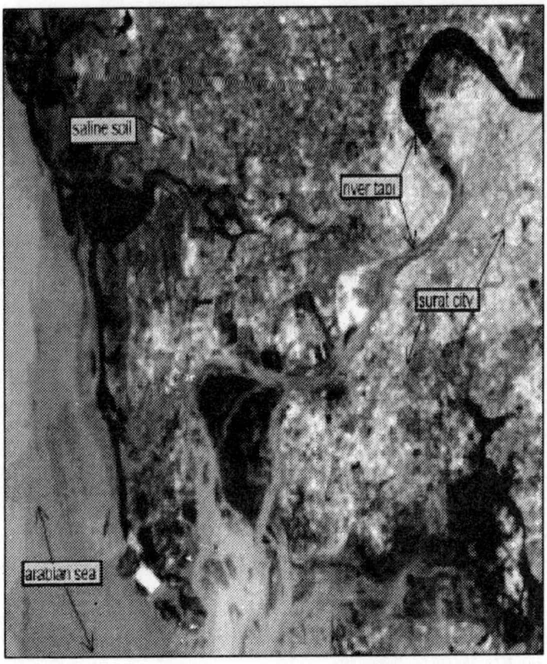

Figure 7.3: Satellite Image of Surat City (*Source*: NRSA)

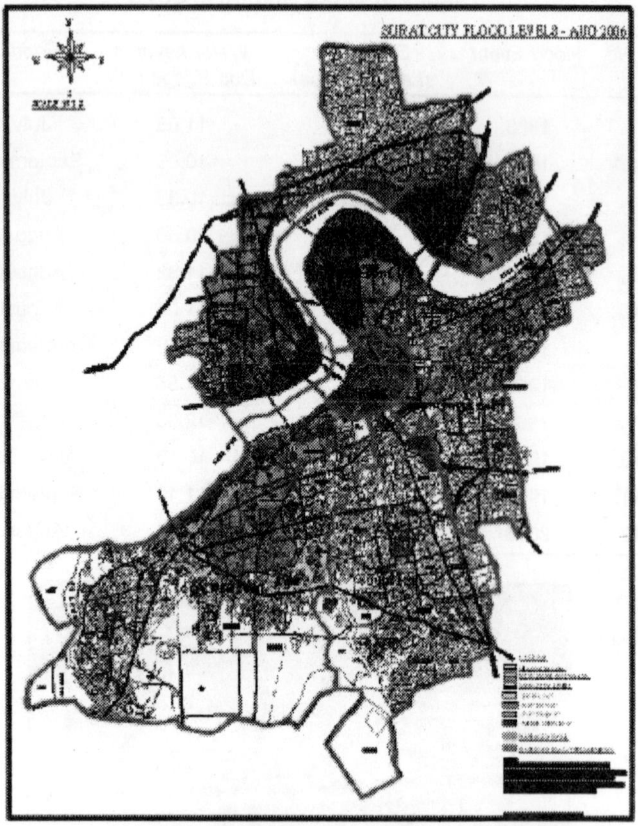

Figure 7.4: Map Indicating Flood Depth at Surat City (Source: S.M.C.)

about the flood through the media. Friends, relatives and neighbours were of big help in warning people. A significant share making up for 20 per cent of respondents said that they got information about the flood either through their friends, relatives or neighbours. The authenticity of message received other than local authority and the media is to be checked. It may cause unnecessary fear and disorder among the community. It is necessary to develop control techniques for flood as well as better warning system if flood occurs. Photographs of a rescue work, flooded river Tapi and area of city are shown in Figures 7.5–7.7.

Figure 7.5: Flooded Area of Surat City

Figure 7.6: A Rescue Work

Various Data Pertaining to the Flood-2006 (*Source*: S.M.C. Office)

The inflow and outflow of water and water levels at Ukai dam are given in Table 7.3 during 01/08/2006 to 20/08/2006. The water

Figure 7.7: Overflowing Bridges during Flood

levels at weir cum causeway and Hop-Bridge at Surat city are also given in the table for the same time and date. The measurements were taken at 12:00 pm on respective dates.

Table 7.3: Data of Flood 2006

Date	*Level at Ukai (Ft.)*	*Inflow at Ukai (Cusec.)*	*Outflow from Ukai (Cusec.)*	*Water Level at Weir Cum Causeway (m)*	*Water Level at Hop-Bridge (m)*
	Danger Level 345ft.			*Danger Level 6 m*	*Danger Level 9.5 m*
01/08/06	331.54	62903	1200	6.71	2.3
02/08/06	333.09	75263	1200	6.88	2.36
03/08/06	334.44	73715	23784	7.20	1.9
04/08/06	335.06	73413	23680	7.02	2.2
05/08/06	335.42	150047	125464	6.97	0.9
06/08/06	337.12	503027	352056	9.70	6.9
07/08/06	342.98	1166390	829829	-	11.90
08/08/06	345.65	922629	908994	-	12.40

Contd...

Table 7.3–Contd...

Date	*Level at Ukai (Ft.)* *Danger Level 345ft.*	*Inflow at Ukai (Cusec.)*	*Outflow from Ukai (Cusec.)*	*Water Level at Weir Cum Causeway (m)* *Danger Level 6 m*	*Water Level at Hop-Bridge (m)* *Danger Level 9.5 m*
09/08/06	346.02	719575	650417	-	12.20
10/08/06	345.17	363632	544795	-	11.25
11/08/06	344.45	234355	299785	-	8.90
12/08/06	344.78	209250	300775	-	6.50
13/08/06	342.03	141817	297203	10.50	7.90
14/08/06	338.50	134318	312168	10.57	8.00
15/08/06	336.14	122704	147916	9.50	6.92
16/08/06	335.32	102292	102292	8.98	5.20
17/08/06	335.16	89335	100891	8.00	3.00
18/08/06	334.23	100868	200232	8.75	3.00
19/08/06	332.37	132824	194527	9.30	5.70
20/08/06	331.00	51619	34824	9.30	5.50

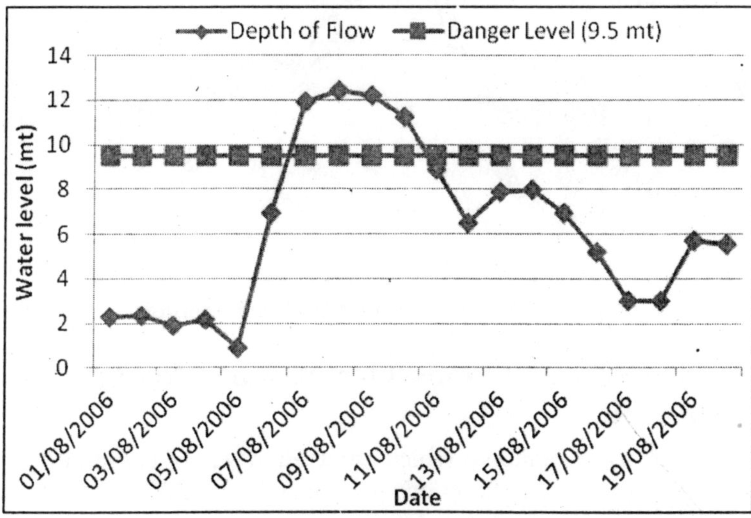

Figure 7.8: Water Level at Hop Bridge, Surat

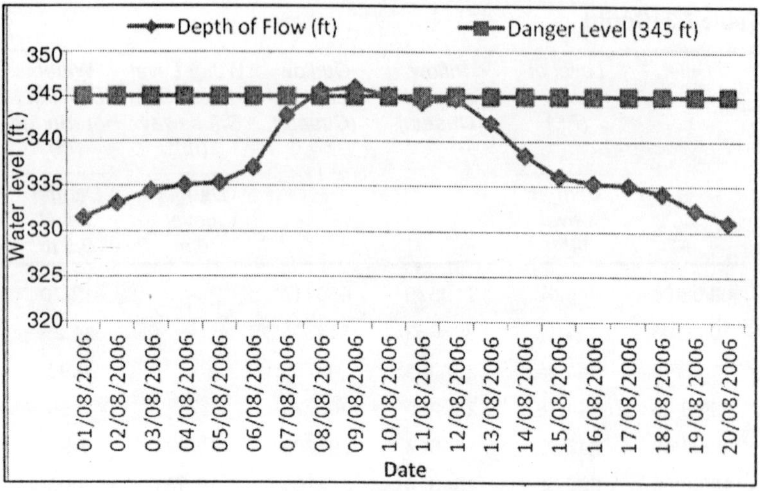

Figure 7.9: Water Level at Ukai Reservoir

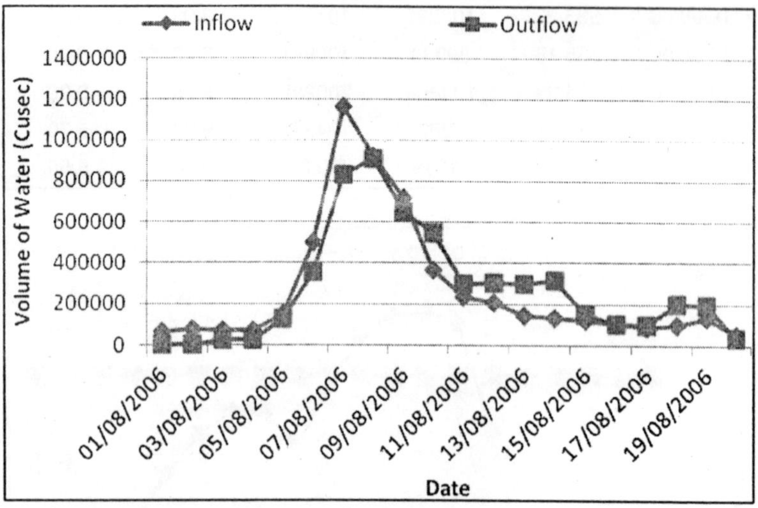

Figure 7.10: Inflow/Outflow at Ukai Reservoir

Comparison of Floods Occured at Mumbai and Surat City

Comparison of Floods which occurred in year of 1998 and 2006 in Surat city with the flood occurred in year of 2005 in Mumbai which is shown in the Table 7.4.

Table 7.4: Comparison of Floods Occurred at Mumbai and Surat City

Sl.No.	Details	Mumbai	Surat	
		Year 2005	Year 1998	Year2006
1.	Total City Area	437sq.km(includes Mumbai city and Suburbs)	112Sq.km	334Sq.km
2.	Total City Population	1,19,14,393 (As per Census 2001) Approx 1.4 crore in July 2005	14.98 Lakhs (As per Census 1991)	~28 Lakhs (As per Census 2001) Approx 40 lakh in 2006
3.	Area under submergence	96 Sq.Km	23.5 Sq. Km	128 Sq.Km
4.	No of Days of submergence	One day	2 Days	7 Days
5.	Population affected	45 Lakh	4 Lakh	25 Lakh
6.	NO of persons shifted to relief centers	25373	63115	63650
7.	Total No. of deaths	450	20	140
	No of death during flood	203		
	No of death during post flood disease	247		
8.	No of people rescued by Army, Paramilitary forces, Police, Fire brigade, etc.	4350	1958	25831
9.	Army, Air force, Coastguard, BSF, Paramilitary forces, etc. deployed for rescue and relief operations	Army, Air force and Navy		Army, Air force, Army Aviation, Coastguard, CRPF 617

Contd...

Table 7.4–Contd...

Sl.No.	*Details*	*Mumbai*	*Surat*	
		Year 2005	*Year 1998*	*Year2006*
10.	No of days of rescue operations	1 day	2 days	5 days
11.	No of food packets supplied	Approx. 2 lakh	Approx. 3.15 Lakh	Approx. 20.5 lakh
12.	No of water pouches and bottles supplied	Approx. 2 lakh	Approx. 35 thousand	Approx. 5.5 lakh
13.	Infrastructure Affected			
	i) Power supply	12 to 15%		95%
	ii) Water supply	8 to 10%	50%	100%
	iii) Drainage	12 to 15%	50%	95%
14.	No of schools closed	All closed on 27th July 2005. 20 to 25 schools were closed for 3 to 4 days. Rests of the schools were opened soon.		All schools
15.	i) Power supply restored	80% within 48 hours		70% within 24 hrs. 85% within 48 hrs. 95% within 72 hrs. 100% within 5 Days
	ii) Water supply restored	70% within 72 hours and 100% within 9 days	100% within 24 Hrs.	22% within 24 hrs. 100% within 48 hrs.

Contd...

Table 7.4–Contd...

Sl.No.	Details	Mumbai	Surat	
		Year 2005	Year 1998	Year2006
	iii) Drainage restored	50% within 24 hours and 100% within 5 days	60% within 24 hrs. 80% within 48 hrs. 100% within 72 hrs.	
16.	Schools Reopened	80% within 24 hrs. and 100% within 48 hrs.		100% after 12 days
17.	Drinking water supplied through tankers	50 tanker and 631 trips (26th July to 6th August)	960 trips	4854 trips 44 million litres.
		Cleaning and rehabilitation work		
18.	Labour deployed (for clean-up operations, post flood)	6600 per shift	6343 per shift	10616 per shift
19.	Solid waste disposed in metric tones	505936(29/7/05 to 16/9/05) 10000 tones against daily avg. of 5000 tones	38888	385537(10/8/05 to 4/9/05) 15000 tones against daily avg. of 1000 tones
20.	Total dead animals disposedSmallBig	16307 (in 3 days)130715000	23902089301	6017(in 4 days)34592558
21.	Machineries deployed for waste removal	JCBs: 107 Dumpers: 438 Compactors: 511	Dumper: 343 JCB: 45	Dumper: 1176 JCB: 223

Contd...

Table 7.4–Contd...

Sl.No.	*Details*	*Mumbai*	*Surat*	
		Year 2005	*Year 1998*	*Year2006*
		Health and Hygiene		
22.	Insecticides used			
	i) Powder in metric tones insecticides used *ii)* Liquid in kiloliters	23MT bleaching powder 800kg carbon phenol	300MT	4740MT
23.	No of medical teams for surveillance	133		
24.	No of people medically examined	3,69,426		
25.	Increase in diseases post flood (one month after flood water receded)	20029 patients		
		Assistance to affected people		
26.	Assistance provided			
	i) Death compensation	Rs. 237.5 lakh 418 persons		Rs. 26 lakh
	ii) Cash doles	Rs. 7657 lakh 824185 persons		Rs. 955.36 lakh Rs. 4052.16 lakh
	iii) Compensation towards loss of household kit	127484 quintal food grains to 98822 families		(3,32,013 families)
	iv) Compensation towards house damage	Nil		

Preventive Remedial Measures

The general remedial measures include preventive as well as curative steps for the control of flood at Surat. Some preventive measures are indicated below.

Reservoir Operation Policy

Full reservoir level at dam is 345 ft., but the water can be stored up to 351 ft. In critical condition water can be stored up to 348.5 ft. safely according to a study report.

Warning System from Upstream

It is necessary to place advanced and computerized rain gauge stations in the catchment area and must be linked with dam site and flood forecasting cell.

Release of Water from the Dam

The release of water from the dam must be in accordance with the water level in the reservoir without affecting the main purpose of irrigation. Superintendent Engineer may be empowered for the same. The information regarding release of water from dam in critical condition must be given to downstream area effectively and timely.

Curative Measures to Control Flood

A flood reduction plan has been prepared which mainly include the curative measures. The details are described below:

De-silting from Reservoir

Due to silting in the reservoir the dead storage capacity of dam has been reduced. The average rate of silting is 1 ft. per year. It is difficult to completely remove mud and silt from the reservoir, however silt removal devices can be installed to minimize the silting effects.

Increasing the Carrying Capacity of River Tapi

The effective waterway of river Tapi has been reduced over years due to silting. The dredging of river in certain reaches and other suitable measures can be implemented to increase the carrying capacity of the river.

Storm Water Drainage Line Parallel to river

Construction of storm water drainage line parallel to river Tapi may be implemented on both the banks and necessary floodgates must be placed.

Construction of Flood Protection Wall

It is necessary to provide flood protection wall on both the banks of river Tapi to protect the residential areas. The existing flood protection wall may not be sufficient in respect of length and height as well as it is in a damaged condition.

Recharge of Groundwater

It is necessary to construct widely spread groundwater recharge wells after proper planning and design in the flood affected area, which can absorb some portion of the flood water. The recharged water will also be useful for increasing the groundwater storage, improving the quality of groundwater and control of sea water intrusion.

Spreading of Flood Water on Saline Soil

It is also proposed to divert the floodwater to the existing saline soil areas outside the city. Such a diversion is possible by construction of suitable diversion channels. The diverted water will serve the purpose of reclamation of saline soil through leaching.

Hi-tech Disaster Management

This can be done by as under:

- ☆ A preparation of counter map for the area from ukai to Sultanabad including Surat city using GPS technique.
- ☆ A preparation of map showing markings of H.F.L as well as their reference (latitude and longitude). This must be co related with discharge and water level at dam.
- ☆ Reference map (co-ordinate) of place should be with rescue and relief team, so that if flood occurs, relief work can be done effectively and efficiently by boat, helicopter etc.
- ☆ Flood hazard map should be prepared using GIS technique.

Diversion of Flood Water to Other Rivers

If the flow in the main channel is to be reduced, a branch channel is sometimes, taken off from the main channel, thus diverting a part of main river water into this branch channel. Under certain critical stages, this method proves to be very effective to relieve the flood strain, and at the same time it assures greater safety to dikes and other flood control structures. It is advisable that the diversion of flow be controlled by providing a regulator, so as to avoid the deterioration of the river in future. The flow of river Tapi during flood can be diverted to other river through interlinking canals, so that the excess water during flood can be managed efficiently. There are five possibility of diverting the flood water from river Tapi.

- ☆ In Purna river near Navsari.
- ☆ In Mindhola river near Sachin.
- ☆ In Kim river near Kosamba.
- ☆ In Sena creek near Dandi, Olpad
- ☆ In Tena creek near KRIBHCO.

Futurastic Development to Control the Flood

- ☆ Varaacha Creek Flood Control And Diversion Scheme
- ☆ The SMC has decided to go ahead with this long-pending project after receiving the irrigation department's advice that the rubber barrage structure was most suited for the city to maximize water conservation, to maximize ground water recharging, to prevent pollution in water and to prevent salinity intrusion in ground water

Conclusion

Surat city (India), situated at the tail end of the rive Tapi is subjected to moderate to heavy floods frequently due to heavy rainfall in the catchment area. Comparison and Effect of various floods in the city have been discussed. Study of the floods at Surat city has been made in this paper also some preventive measures, curative measures have been suggested. Also the futuristic development plan which can minimise the flood by SMC is given. To prevent flood in future proper management of gate opening and closing, proper schedule and rule level should be maintain. Proper disaster management should be planned to stand against such a natural disaster.

References

Acharya, Akash, Das, Biswaroop and Pandya, Kiran, *et al.*, 2006. "Surat 2006 Floods: A Citizen's Report" Centre for Social Studies (CSS) Surat and Department of Human Resource Development (DHRD), Veer Narmad South Gujarat University, Surat, October.

Kumar, Ajay, 1995. *Flood Control and Settlement Planning*. Mohit Publication, New Delhi.

Patel, J.N., 2002. Flood management over a region of river Tapi using GPS. *Journal of Geospatial Today*, 1(4) October–November.

Shah, Jagruti, 2008. *Seminar Report on Case Study of Surat Flood, 2006.* GEC, Surat, India.

Shelat, R.N., 1969. *Flood Control and the Use of River Water Resources.* South Gujarat University, Surat.

Websites

http://www.indiatogether.org/environment/interlink.htm

http://www.sandrp.in/new/surat floods aug08.pdf.com

http://www.gpsdotnet.com

http://www.gislounge.com/ll/hydrology.shtml

http://www.suratmunicipal.org

Chapter 8

Impacts of Land Use and Climate Change Scenarios on the Water Flux: A Case Study of Sal River Watershed, Chamba, H.P.

☆ *Kuldeep Pareta and K.K. Mahapatra*

ABSTRACT

With the flow of the Sal River becoming increasingly erratic especially in the upper reaches, attention has been directed to land use change as the major cause of this problem. The semi-distributed hydrological model soil and water assessment tool (SWAT), IRS-P6-LISS-IV-Mx (5.8m) and IRS-P6-LISS-III (23.5m) satellite imageries were utilized in the Sal River watershed in order to (*i*) map existing field scale land use practices in order to determine their impact (*ii*) determine the impacts of land use change on water flux; and (*iii*) determine the impacts of rainfall and temperature variations based on the Intergovernmental Panel on Climate Change (IPCC) projections on the water flux of the Sal River.

This study found that the different scenarios are affecting the water-balance components differently. Land use changes resulted in a slightly more erratic discharge while rainfall and temperature changes had a more predictable impact on the discharge and water-balance components. These findings demonstrate that the model results show the flow was more sensitive to the rainfall changes than land use changes. It was also shown that land use changes could reduce dry season flow, which is the most important problem in the watershed. The model shows deforestation in the Binnu, Suri, and Mills forest increased the peak flows, which can also lead to high sediment loading in the Sal River. The effect of the land use and climate-change-scenarios on the sediment and water-quality of the river needs a thorough understanding of the sediment transport processes in addition to observed sediment and water-quality data for validation of modeling results.

Keywords: *Land use, Climate change, SWAT model, Water flux, Remote sensing.*

Introduction

Water is the most precious gift of nature to the mankind and has been recognized as one of the most vital natural resources. It is not so just, because water sustains life but it is a renewable resource. It is also the most essential input for agriculture industry and power generation. The availability of water with proper quality and quantity at appropriate time and space are of great importance. The role of water resource development in the overall economic development of the country is well recognized. According to Reddy (1972) agriculture contributes thirty percent of India's Gross National Profit (GNP) and 60 per cent of the employment potential, over two third of the country's population is agriculture dependent and irrigation contributes to 55 per cent of agricultural output. The water management is very essential to maintain quality, quantity and the availability of water due to increase in population, rapid urbanization and industrial growth. Though its availability is limited, yet demand for water is ever increasing. It has direct impact on human being and socio-economic development. Hence, the need of proper planning management of the precious resource has become the matter of utmost urgency.

Remote Sensing is perhaps most aptly defined by B.F. Richason as the science technology and skill of "imaging from orbital and sub-orbital altitudes, in various (and simultaneously in several) parts of the visible and invisible (EMR) spectrum and converting data into information by visual and digital processing". Remote Sensing, particularly from continually revisiting robotized satellites, provides a very effective supplemental source and technique of updated field observation and mapping of not only tangible phenomena but also of other elements that can be deduced from characteristics thereof. India is one of the major global players in several fields of space technology, including satellite remote sensing. Its current IRS series satellites are currently providing high-resolution multi spectral data. With our own remote sensing satellites and earth station in place, it becomes the economic and fastest technique of extensively surveying and mapping inaccessible areas.

The specific **objectives** of this study were to: map existing land use practices using remote sensing and field observations, determine the impacts of land use change, rainfall and air temperature variation on the water flux of the Sal River in Chamba. The findings of this study provided scenarios on the impacts of land use and climate change in the Sal River Watershed therefore adding to the existing literature and knowledge base with a view of promoting better land use management practices in Ravi River Basin, Himachal Pradesh and application of the SWAT model in similar densely populated, highly agricultural watersheds all over the world.

Methods

For the SWAT model application in the Sal Basin, detailed dataset inputs had to be prepared. These datasets included detailed land use/land cover map, soil classification map, and climate data on a daily time-step. The land use map is an important input for the model and involved analysis of remote sensor data in order to generate a detailed accurate map for use in the SWAT model.

Study Area

Sal River is a moderate size northern sub-tributary of the river Ravi, and originating at the eastern part of Nichla Supeka, Chamba district, it is flow essentially NE to SW and over 32.3 km to join the Ravi River near Chamba town of Himachal Pradesh. The Sal River Watershed (Figure 8.1) is shared Chamba Tehsil and is located in North-East Chamba between longitudes 76° 7' 17.36" E and 76° 22' 56.3" E, latitudes 32° 31' 45.2" N and 32° 41' 21.2" N. It covers about 265.41 km^2 and is characterized by different types of land cover and land use as a result of different human activities carried out by the

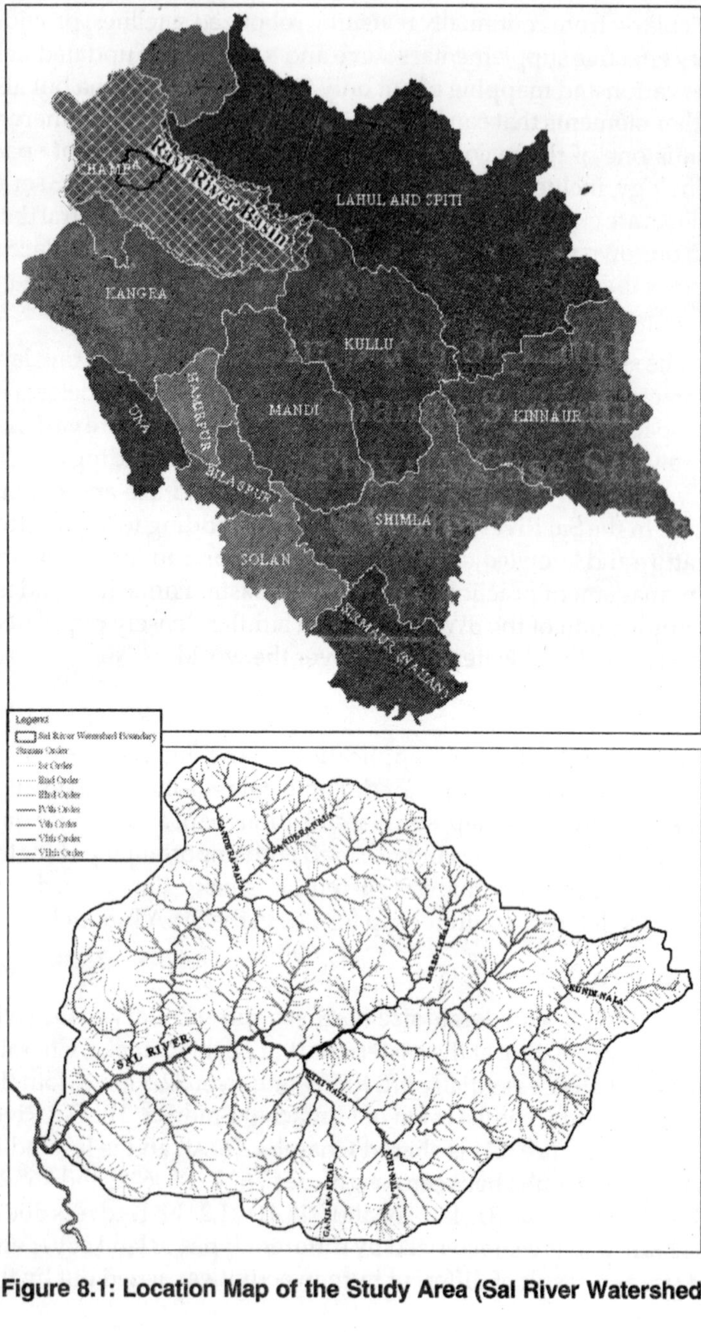

Figure 8.1: Location Map of the Study Area (Sal River Watershed)

stakeholders in various parts of the watershed. The total length of the Sal River is 32.3 km. Though there is no main tributaries of the Sal River, there are some small tributaries pouring into the river, notable amongst there are Gandera Nala, Sanluta Nala, Saredi Khad on the left bank, and Kundi Nala, Kiri Nala, Ganji Ka Khad, and Sarotha Nala on the right bank.

Swat Model Description

The Soil and Water Assessment Tool (SWAT) is a hydrological model that can be applied at the river basin, or watershed scale. It was developed for the purpose of simulation of impact of land management practices on water, and sediment yields in large watersheds with varying soils, land use and agricultural conditions over extended time periods (Neitsch *et al.*, 2005). Arnold *et al.* (1998) defines SWAT as a semi-distributed, time continuous simulator operating on a daily time step. It is developed for assessment of the impact of management and climate on water supplies, and sediment yields in sub-basins and larger basins. The program is provided with an interface in ArcGIS (Di Luzio *et al.*, 2002) for the definition of watershed hydrologic features and storage, as well as the organization and manipulation of the related spatial and tabular data.

Land Use Data Classification

The SWAT model requires a spatially explicit land use map as an input in order to simulate the hydrology of a watershed. Land use data was obtained by the classification of remote sensor data specifically, satellite imagery from the IRS-P6 LISS-IV Mx (5.8 m spatial resolution) and IRS-P6 LISS-III (23.5 m spatial resolution) sensor built for earth observation purposes, and it suitable for land cover classification (Van der Meer *et al.*, 2002). The LISS-III image of Path 94, Row 47 from the 14th December 2010, and LISS-IV Mx image of Path 33334, Row 032 from 2nd December, 2010 were selected for the classification.

Visual image interpretation is defined as the act of examining images to identify objects and judge their significance. An interpreter studies remotely sensed data and attempts through logical process to detect, identify, measure and evaluate the significance of environmental and cultural objects, patterns and spatial relationships. It is an information extraction process. The visual land use classification system was used to classify the satellite imagery of the study area, which are settlement, agricultural, reserve forest, protect forest, jungle, scrub land, water bodies, and river.

A land use/land cover classification scheme was formulated that would accurately and adequately represent the land cover/land use within the Sal River watershed (Table 8.1). This scheme however follows the basic principles of the USGS Land use/land cover classification system (LULCCS) for use with remote sensor data level classification (Anderson *et al.*, 1976).

Table 8.1: Land Use/Land Cover Type Reclassification into SWAT LU/LC Classes

Sl.No.	*Land Cover Type*	*Area (in Hectare)*	*SWAT LU/LC Type*
1.	Settlement	383.82	Settlement
2.	Agricultural	4586.61	Agricultural Land Generic
			Agricultural Land Close Grown
3.	Reserve Forest	999.60	Forest Evergreen
4.	Protect Forest	12574.29	Forest Deciduous
5.	Jungle	508.69	Forest Mixed
6.	Scrub Land	7376.43	Range Grasses
7.	Water Bodies	00.00	Water
8.	River	133.62	Water
	Total	**26563.06**	

Hydrological Modeling

The SWAT (Soil and Water Assessment Tool) model was the ideal choice for use in this study because of various reasons; it is a physically based model that requires specific information about weather, soil properties, topography, vegetation and land management practices which it uses as inputs to simulate the physical processes associated with water movement, nutrient transport, crop growth and sediment movement. This enables it to model ungagged watersheds and more importantly, quantify the impact of alternative input data such as changes in land use, land management practices and climate on water quality and quantity. Secondly, it uses readily available data, while more inputs can be used to simulate more specialized processes and it is still able to operate on minimum data, which is an advantage especially when working in areas with insufficient or unreliable data. Third, the SWAT model is computationally efficient, able to run simulations of very large basins or management practices without consuming large amounts of time and expenses. Lastly, it is a continuous time or a long-term yield model able to simulate long-term impacts of land use, land management practices and build-up of pollutants (Neitsch *et al.*, 2005). These qualities of the SWAT model enabled the quantification of long-term impacts of land use changes, variations in rainfall and air temperature on the hydrology of the Sal River watershed.

DEM

The digital elevation model (DEM) of 30m by 30m resolution for the study area obtained from the ASTER was used. The DEM (Figure 8.2) gives the elevation of a particular point at a particular spatial resolution and was used in the delineation of the watershed and analysis of the land surface characteristics and drainage patterns.

Soil Data Classification

Soil data was obtained from the *National Bureau of Soil Survey and Land Use Planning, Indian Council of Agricultural Research, Nagpur (NBSS and LUP), and Department of Agricultural, Government of Himachal Pradesh, Shimla*. GIS layers were obtained and used in the hydrological model as one of the main inputs to the SWAT model which requires soil property data such as the texture, chemical composition, physical properties, available moisture content, hydraulic conductivity, bulk density and organic carbon content for the different layers of each soil type (Setegn, 2008). A user table specific for the Sal River watershed soil layer (Figure 8.3) was appended to the soil table in the SWAT database by using Arc Toolbox in ArcGIS.

Land Use

Land use data for the year 2010 was obtained by analysis of IRS-P6 Resourcesat-1 LISS-III (Figure 8.4) and LISS-IV Mx (Figure 8.5) satellite imageries in the process described previously in this chapter resulting in land cover maps of the study area. These land cover maps were then converted to shapefiles and aggregated to make them easier to input into the model for use in the hydrological modeling exercise. Land use and management is an important factor

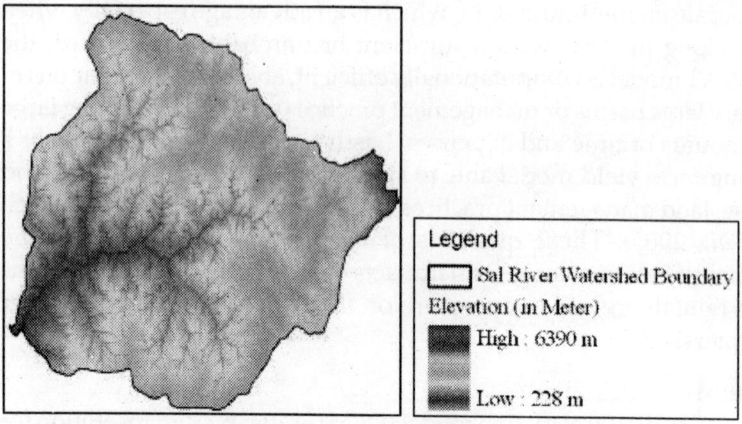

Figure 8.2: DEM (30m)

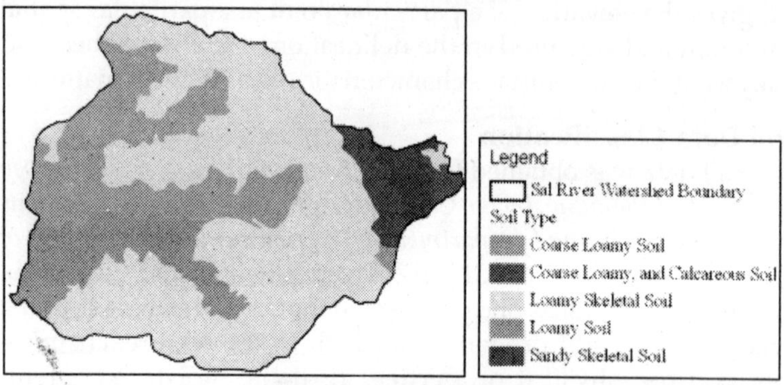

Figure 8.3: Soil Type

affecting different processes in the watershed such as surface runoff, erosion and evapotranspiration. Reclassification of the land use map (Figure 8.6) was done in order to present them in a form that is acceptable in the model and this is the USGS Land use/Land cover classification scheme for Use with Remote Sensor data level classification (Anderson *et al.*, 1976).

Climate Data

The study area receives precipitation due to the South-West Monsoon as well as the Western disturbances that pass over the North-West part of the country during winter. There are thirteen non-recording type rain gauge stations in the catchment area of the

Figure 8.4: IRS P6 LISS-III Data (23.5m)

Figure 8.5: IRS P6 LISS-IV Data (5.8m)

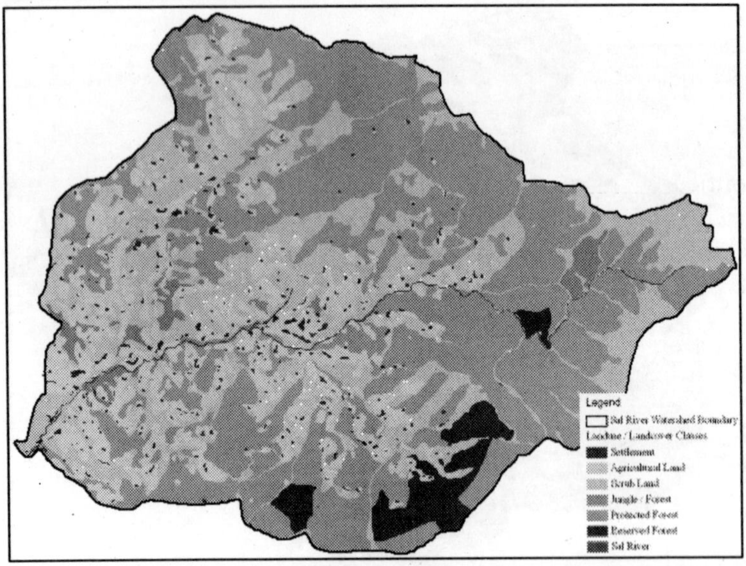

Figure 8.6: Land Use/Land Cover Map-2010

river Ravi. The normal annual rainfall has been recorded in millimeters at all these stations for the period 2006 to 2010 by the office of *Director Land Records, Revenue Department (HP).*

Climate data used in the SWAT model consists of daily rainfall, temperature, wind speed, humidity and evapotranspiration data. The temperatures in summer vary between 38 °C (100 °F) and 15 °C (59 °F) and in winter; 15 °C (59 °F) and 0 °C (32 °F). The maximum temperature recorded in summer is 39 °C (102 °F) and the minimum temperature in winter is -1 °C (30.2 °F). The average annual rainfall in the town is 785.84 millimeters (30.939 in). The climate data was able to provide continuous and complete data ranging from the years 2006 to 2010, which was used in the model simulations.

Table 8.2: Rainfall Data of the Study Area (Sal River Watershed)

Sl.No.	*District*	*2005*	*2006*	*2007*	*2008*	*2009*	*2010*
1.	Chamba (mm)	1952.7	1583.7	706.4	857.2		1100.3

Source: Directorate of Land Records, Revenue Department, H.P.

Table 8.3: Sensitivity Ranking of Parameters Towards Water Flow

Sensitivity Rank	*Sal Watershed Area Rain Gauge*	*Sal REF*
1	ESCO	CN2
2	CN2	GWQMN
3	ALPHA-BF	CN2
4	GWQMN	SOL-Z
5	SOL-Z	ALPHA-BF
6	REVAPMN	SOL-AWC
7	SOL-AWC	REVAPMN
8	CH-K2	CANMX
9	BLAI	GW-REVAP
10	CANMX	BLAI

ESCO: Soil Evaporation Compensation Factor; CN2: Initial SCS runoff Curve Number for Moisture Condition II; GWQMIN: Threshold depth of water in the shallow aquifer for return flow to occur (mm H_2O); SOL-Z: Depth to bottom of soil layer (mm); ALPHA-BF: Base Flow alpha factor (days); REVAPMN: Threshold depth of water in the shallow aquifer for "revap" to occur (mm H_2O); SOL-AWC: Available water capacity of soil layer (mm/mm); CANMX: Maximum canopy storage (mm H_2O); BLAI: Maximum potential leaf area index for land cover/plant; GW-REVAP: Groundwater "revap" coefficient.

Table 8.4: Model Evaluation Statistics for Daily Discharge (Sal River)

Statistic	*RFE*		*Rain Gauge*	
	Cal	*Val*	*Cal*	*Val*
NSE	0.485	0.0807	-0.445	0.0178
R^2	0.530	0.233	0.072	0.257
R	0.728	0.483	0.0269	0.507

Table 8.5: Model Evaluation Statistics for Monthly Discharge (Sal River)

Statistic	*RFE*		*Rain Gauge*	
	Cal	*Val*	*Cal*	*Val*
NSE	0.586	0.094	-0.533	-0.057
R^2	0.645	0.325	0.085	0.321
R	0.803	0.570	0.291	0.566

River Discharge

In the Sal River watershed, collected the discharge data for the year 2006 and for the rain gauge model, 2 years were used for calibration and 2 years were used for validating the model. For the RFE model, 3 years were used for the calibration and 2 years for validation of the model. The length of the simulations was determined by the availability and length of time series data for discharge, air temperature and rainfall, which are key pieces in the model simulation.

The discharge of the Sal and its major snow fed tributaries dwindles down to lowest during winter months, from December to February and starts increasing from February end due to the melting of snow. The bulk contribution is from rainfall in the monsoon months.

Table 8.6: Hydrological Characteristics of the Study Area

Sl.No.	*Characteristics*	*Units*
1.	Total Catchment Area	26563.06 Hectare
2.	Snow fed Area	13 Hectare
3.	Rain fed Area	26550.06 Hectare
4.	River	Sal Nala
5.	Average Annual Rainfall	1100 mm
6.	Average Annual Runoff	6960 mcum
7.	Average Annual Sediment Load	6.8 mcum
8.	Sediment Yield	0.144 Ha. M./Sq. Km/Year
9.	River Discharge	89.67 to 71.74 Cum/Sec

Model Run

To set up a hydrological SWAT model, basic data are required: topography, soil, land use and climatic data (Schuol *et al.*, 2006). The model setup involved five steps: (1) data preparation, (2) sub-basin discretization, (3) HRU definition, (4) parameter sensitivity analysis, (5) calibration and uncertainty analysis.

The DEM was projected to the required projection parameter, which is UTM Zone 43 North. A mask was used to reduce the area for stream delineation and analysis of terrain drainage patterns of the land surface. The streams were delineated from the DEM, which accurately captured their true location on the ground. The land use/ land cover layer was reclassified into the SWAT/USGS land use code as per required by the model and linked to a user table with the land use code. Watershed and sub-watershed delineation was carried out using the DEM and has various steps including DEM setup, stream definition, outlet and inlet definition, watershed outlets selection and definition and calculation of sub-watershed parameters. The resulting sub-watersheds were then divided into units based on their unique combination of land use, soils and slope combinations and these units are known as HRUs (hydrologic response units). The model was run on a default simulation of 5 years from 2006 to 2010 for the Rain gauge data and from 2006 to 2010 a period of five years for the RFE data.

Scenario Analysis

Land Use Scenarios

To explore the sensitivity of SWAT outputs to land use and the effect of land use/land cover changes on the discharge of the Sal Rivers, land use scenarios were explored. Attention was paid to ensure these were realistic scenarios in accordance to the ongoing trends of land use change within the study area. The percent coverage and details of the conversions are presented in the Table 8.7. The land use scenarios included:

Partial Deforestation

This scenario-involved manipulation of the forest cover reducing it partially by converting the deciduous forest type to small scale or close grown agricultural land.

Complete Deforestation

This scenario involved replacing all the existing forest cover with grassland to simulate a complete absence of forest cover in the watershed.

Conversion of Forest to Agriculture

Replacement of forest land by agriculture is a common trend within the study area and is seen to be one of the major causes of erratic river flows and increased sediment load in the Sal Rivers. This scenario was carried out by replacing all forest cover with agriculture particularly small-scale agriculture.

Table 8.7: Per cent Areal Coverage of Land Use/Land Cover–2010

Land Use Scenario/ Watershed	*Landuse 2010*	*PD 2010*	*CD 2010*	*FA 2010*	*Sal Watershed*
Settlement	383.82	0	0	0	383.82
Agricultural Land Generic	2751.966	0	0	0	2751.966
Agricultural Land Close Grown	1834.644	0	0	0	1834.644
Forest Evergreen	999.60	723	56	220.6	999.60
Forest Deciduous	12574.29	9786	674	2114.29	12574.29
Forest Mixed	508.69	413	33	62.69	508.69
Range Grasses	7376.43	6345	397	634.43	7376.43
Water	133.62	0	0	0	133.62

PD is Partial Deforestation, CD is Complete Deforestation, FA is conversion of forest to Agricultural

Climate Change Scenarios

According to the Intergovernmental Panel on Climate Change (2007), climate change can be defined as an identifiable change in the state of the climate by change in the mean and/or variability of its properties and that persists for an extended period, typically decades or longer.

Trends from 2006 to 2010 have been observed in precipitation and have seen a decrease in precipitation in the Chamba district, which has been accounted for in the precipitation reduction scenarios carried out in this study. In the case of projection of future changes of changes in climate for the 21st century and beyond, consideration was given to the scenarios described in the *IPCC Special Report on Emission Scenarios* and the scenario chosen for the study region was A1Fl *(rapid economic growth and fossil intensive)*. This was based on the fact that Himachal Pradesh (India) is a developing country with a rapid population growth, is embracing new technology and is dependent on fossil fuel as its major source of energy. The A1Fl scenario projects a temperature increase with a best estimate of 4.0 degrees centigrade with a likely range of between 2.4–6.4 degrees centigrade. Precipitation in the 21st century and beyond is projected to increase by between 5 and 20 per cent for the region. This prompted the precipitation and temperature scenarios below purposely set to capture the effect of both increase and decrease of precipitation and increase in surface temperature.

The different climate scenarios explored included:

1. Rainfall scenarios: 0 per cent, ±10 per cent and ±20 per cent
2. Temperature scenarios: 0 per cent and +5 per cent
3. Combination of rainfall and temperature scenarios

These were carried out by replacing the precipitation and temperature files in the model and running the simulations with the best parameters acquired from the calibration process.

Results and Discussion

The results of this study are divided into two categories: land cover classification and hydrological modeling. The land cover mapping provides data on the type of land use/land cover types present within the Sal River watersheds. The hydrological modeling section provides data on the discharge amounts and water balance components of the Sal River watersheds and the effect of land use and climate change.

Land Cover Mapping

The land use map was reclassified into SWAT land use/land cover classes (Table 8.1) to be used in the hydrological modeling process.

Hydrological Modeling

A sensitivity analysis was carried out and the 10 most sensitive parameters (Table 8.3) were chosen for calibration of the model. The hydrological modeling exercise resulted in discharge simulation values for the Sal watershed for different rainfall inputs; Rain gauge measurements and radar rainfall estimates (RFE). The discharge hydrographs for daily and monthly data were compared for calibration and scenario analysis.

Model Calibration and Validation

Parameter adjustment was carried out in conjunction with the statistical evaluation until an acceptable correlation or resemblance between the two datasets was achieved. Calibration is the process of estimating model parameters by comparison of model predictions or output for a given set of assumed conditions with observed or measured data for the same conditions *(Moriasi et al., 2007)*. Comparison was carried out for the datasets obtained and the resulting statistics for the daily and monthly simulations are shown in the Tables 8.4 and 8.5. Statistics such as the Nash-Sutcliffe Efficiency (NSE), Pearson's correlation coefficient (r) and the Coefficient of Correlation (R2), which were used to describe and compare the different datasets (observed and simulated).

In the case of the Sal Watershed (Chamba) rain gauge data models, there was a clear underperformance of the models in the case of discharge simulation as shown by the different model evaluation statistics in *Moriasi et al. (2007)*. Calibration of the rain gauge data produced NSE values of -0.533 for Sal Watershed (Chamba), which are considered poor. The RFE data on the other hand produced NSE values of 0.586 for the Sal Rivers and was considered good results.

Validation was also carried out for the model simulations and is defined as the process of demonstrating that a given site-specific model is capable of making sufficiently accurate simulations *(Refsgaard, 1997)*. This was carried out to determine whether these models were suitable for evaluating the impact of land use and climate change.

For the RFE models, NSE values of 0.094 was obtained for the Sal Watershed (Chamba), and taking into consideration the errors that may have been introduced by missing data values, the models was considered suitable for predicting the impacts of climate and land use change. According to *Abbaspour et al. (2006)*, watershed scale model calibration is challenging and is impeded by uncertainties like watershed processes unknown to the modeler, processes not captured by the model and simplification of the processes by the model.

Climate Change Scenarios

Assuming accurate estimates of the water balance components, SWAT was used to evaluate the impacts of various scenarios of climate change on Sal rivers. The combined discharge hydrographs for the climate change scenarios shown below help single out the impact a single climate change event however unlikely, would have on the discharge of the Sal river. Table 8.9 shows the percent changes in the annual averages of Sal Watershed water balance components and Table 8.10 shows the ratio of water balance components to precipitation for these climate change scenarios.

Table 8.8: Annual Average Water Balance Components for the Calibrated Sal Watershed Models

Sl.No.	*Components*	*Rain Gauge–2006*	*RFE–2010*
1.	PRECIP (mm)	1583.70	1100.03
2.	SURQ (mm)	17.89	11.53969
3.	LATQ (mm)	72.24	43.20114
4.	GW Q (mm)	422.26	482.4712
5.	REVAP (mm)	26.06	3.488976
6.	DA RCHG (mm)	26.75	25.39533
7.	GW RCHG (mm)	535.19	507.9367
8.	WYLD (mm)	511.20	536.3799
9.	PERC (mm)	535.90	510.8342
10.	ET (mm)	939.57	531.367
11.	PET (mm)	1369.82	1182.041
12.	TLOSS (mm)	1.20	0.822115
13.	SEDYLD (T/HA)	0.81	0.705816

PR–10: Precipitation reduced by 10 per cent; PR–20: Precipitation reduced by 20 per cent; TM+5: Air Temperature increase by 5 per cent; PR+10: Precipitation increase by 10 per cent; PR+20: Precipitation increase by 20 per cent.

Generally, the reduction of precipitation brought about a reduction in available water in the watersheds reducing base flows to very low levels. The increase in temperature also reduces the water availability to some degree by increasing evapotranspiration in the watershed thus reducing amount of water and discharge. According to Ficklin *et al.* (2009), temperature is one of the most important factors governing plant growth and depending on the optimum temperature of the plants; the plant growth cycle will be shifted also affecting the water balance components. Increases in precipitation by 10 percent and 20 percent increased the discharge and base flow in the rivers but on the other hand may have negative effects across land such as erosion and in the reach such as increased sediment load and flooding.

Table 8.9: Per cent Changes in the Annual Averages of Sal Watershed Water Balance Components for Climate Change Scenarios

Components	*PR-10*	*PR-20*	*TM+5*	*PR-10 TM+5*	*PR-20 TM+5*	*PR+10 TM+5*	*PR+20 TM+5*
PRECIP (mm)	–9.92	–19.98	0.00	–9.92	–19.98	10.07	19.98
SURQ (mm)	–34.14	–60.38	–5.04	–38.23	–63.51	37.71	90.79
LATQ (mm)	–14.97	–29.89	–2.32	–17.15	–31.84	12.90	27.80
GW Q (mm)	–17.31	–34.50	–4.09	–21.29	–38.26	13.52	30.76
REVAP (mm)	–35.92	–63.51	–6.61	–41.38	–68.39	34.48	77.30
DA RCHG (mm)	–17.29	–34.50	–4.07	–21.28	–38.26	13.54	30.75
GW RCHG (mm)	–17.31	–34.50	–4.09	–21.29	–38.26	13.52	30.76
WYLD (mm)	–17.47	–34.66	–3.97	–21.31	–38.26	13.97	31.77
PERC (mm)	–17.27	–34.42	–4.08	–21.24	–38.17	13.49	30.69
ET (mm)	–1.94	–4.49	3.91	1-83	–0.98	5.70	7.28
PET (mm)	0.00	0.00	3.15	3.15	3.15	3.15	3.15
TLOSS (mm)	–25.61	–50.00	–1.22	–28.05	–52.44	28.05	62.20
SEDYLD (T/HA)	–39.49	–67.05	–3.69	–43.47	–69.89	50.43	124.15

PD: Partial Deforestation; CD: Complete Deforestation; FA: Forest replaced by Agriculture

Table 8.10: Ratio of Water Balance Components to Precipitation for the Sal Watershed Climate Change Scenarios

Components	*RFE COMP*	*PR-10*	*PR-20*	*TM+5*	*PR-10 TM+5*	*PR-20 TM+5*	*PR+10 TM+5*	*PR+20 TM+5*
PRECIP (mm)	1.000	1.000	1.000	1.000	1.000	1.000	1.000	1.000
SURQ (mm)	0.010	0.008	0.005	0.010	0.007	0.005	0.013	0.017
LATQ (mm)	0.039	0.037	0.034	0.038	0.036	0.033	0.040	0.042
GW Q (mm)	0.439	0.403	0.359	0.421	0.383	0.338	0.452	0.478
REVAP (mm)	0.003	0.002	0.001	0.003	0.002	0.001	0.004	0.005
DP AQ RCHRG (mm)	0.023	0.021	0.019	0.022	0.020	0.018	0.024	0.025
TOTAL AQ RCHRG (mm)	0.462	0.424	0.378	0.443	0.403	0.356	0.476	0.503
WYLD (mm)	0.488	0.447	0.398	0.468	0.426	0.376	0.505	0.536
PERC (mm)	0.464	0.426	0.381	0.445	0.406	0.359	0.479	0.506
ET (mm)	0.483	0.526	0.577	0.502	0.546	0.598	0.464	0.432
PET (mm)	1.075	1.193	1.381	1.108	1.230	1.385	1.007	0.924
TLOSS (mm)	0.001	0.001	0.000	0.001	0.001	0.000	0.001	0.001
SED (T/HA)	0.001	0.000	0.000	0.001	0.000	0.000	0.001	0.001

PD: Partial Deforestation; CD: Complete Deforestation; FA: Forest replaced by Agriculture

Land Use Change Scenarios

The resulting hydrographs show the effect the different land use scenarios had on the river discharge. From observation of the graphs, it is evident that all the land use scenarios significantly reduced the base flow and average flow of the whole period of simulation. The partial deforestation and forest to agriculture scenarios resulted in high peak flows and lower base flows while the complete deforestation scenario was characterized by high peak flows but has a base flow that appears almost equivalent to that of the present day scenario (RFE calibrated model). Details on how these different land use scenarios affected the different water balance components can be seen in Table 8.11.

Combination of Land Use and Climate Change Scenarios

The shown hydrographs display the discharge outputs of the land use-climate change scenarios, which were more realistic scenarios in terms of future projections of land use and climate change. The most plausible land use scenarios in the case of the Sal Watershed are the three covered in this study, with complete deforestation being the least likely to happen among the three. In the case of climate change, the most plausible are the combinations of temperature increase and precipitation increase and decrease depending on geographic location as projected by the IPCC though precipitation reduction is a more often occurrence in the study area as of today and therefore were included in the scenarios analysis.

The resulting hydrographs and annual average for water balance components (Table 8.11) were able to graphically show the effects of the combined scenarios in terms of stream response. From the hydrographs (Figure 8.7 and 8.8) and Figure 8.10, the effect of land use on the discharge hydrographs and water balance components was evident. The conversion of forest to agriculture scenario had the lowest base flows and the cause for this is the reduction in ground water recharge which is shown in Figure 8.10 and Table 8.11 with reductions of up to 49.99 per cent when precipitation is reduced by 20 per cent and a reduction of 32 per cent when precipitation is reduced by 10 per cent and a reduction of 48 per cent at normal precipitation. Complete deforestation on the other hand, saw an increase in surface runoff and this occurred in the 20 per cent precipitation reduction in Sal watersheds.

Table 8.11: Percent Changes in the Annual Averages of Sal Watershed Water Balance Components for Land Use-Climate Change Scenarios

Component	*PD*	*CD*	*FA*	*PD PR-10 M+5*	*PD PR-20 M+5*	*CD PR-10 M+5*	*CD PR-20 M+5*	*FAP R-20 M+5*	*FA PR-20 M+5*
PRECIP (mm)	0.00	–9.92	–19.98	0.00	–9.92	–19.98	0.00	–9.92	–19.98
SURQ (mm)	6.94	–24.08	–48.85	20.58	–13.61	41.47	31.49	–5.99	–36.31
LATQ (mm)	–1.56	–20.65	–35.94	–0.66	–17.55	–32.60	–14.91	–30.31	–43.44
GW Q (mm)	–4.16	–30.49	–49.05	–0.75	–24.38	–43.01	–9.39	–33.28	–51.00
REVAP (mm)	–1.84	–12.76	–26.76	–0.89	–10.03	–23.28	–2.80	–12.76	–27.37
DP RCHRG (mm)	–4.04	–29.68	–48.09	–0.75	–23.79	–42.21	–9.06	–32.33	–50.00
GW RCHRG (mm)	–4.06	–29.68	–48.12	–0.79	–23.78	–42.21	–9.09	–32.35	–49.99
WYLD (mm)	–2.00	–28.50	–47.85	2.93	–21.96	–41.88	–3.00	–28.49	–47.98
PERC (mm)	–4.22	–29.92	–48.36	–1.10	–24.07	–42.49	–9.71	–32.91	–50.50
ET (mm)	1.64	3.46	0.23	–1.93	–1.33	–4.20	2.46	3.50	0.35
PET (mm)	0.00	3.15	3.15	0.00	3.15	3.15	0.00	3.15	3.15
TLOSS (mm)	9.90	–10.15	–29.21	23.76	1.49	–19.80	45.30	20.30	–4.46
SED (T/HA)	22.39	–12.37	–42.73	11.90	–22.78	–50.15	41.48	1.19	–33.21

Table 8.12: Ratio of Water Balance Components to Precipitation for the Sal Watershed Land Use-Climate Change Scenarios

Component	*DEF*	*PD*	*CD*	*FA*	*PD PR-10 TM+5*	*CD PR-10 TM+5*	*FA PR-10 TM+5*	*PD PR-20 TM+5*	*CD PR-20 TM+5*	*FA PR-20 M+5*
PRECIP (mm)	1.000	1.000	1.000	1.000	1.000	1.000	1.000	1.000	1.000	1.000
SURQ (mm)	0.072	0.077	0.087	0.094	0.061	0.069	0.075	0.046	0.053	0.057
LATQ (mm)	0.040	0.040	0.040	0.034	0.036	0.037	0.031	0.032	0.034	0.029
GW Q (mm)	0.285	0.273	0.282	0.258	0.220	0.239	0.211	0.181	0.203	0.174
REVAP (mm)	0.013	0.013	0.013	0.013	0.013	0.013	0.013	0.013	0.013	0.012
DP AQ RCHRG mm)	0.016	0.015	0.016	0.014	0.012	0.013	0.012	0.010	0.011	0.010
Total AQ RCHRG mm)	0.316	0.303	0.313	0.287	0.246	0.267	0.237	0.205	0.228	0.197
WYLD (mm)	0393	0.385	0.405	0.381	0.312	0.341	0.312	0.256	0.286	0.256
PERC (mm)	0.315	0.301	0.311	0.284	0.245	0.265	0.234	0.203	0.226	0.195
ET (mm)	0.567	0.576	0.556	0.581	0.651	0.621	0.651	0.710	0.679	0.711
PET (mm)	1.075	1.075	1.075	1.075	1.230	1.230	1.230	1.385	1.385	1.385
TLOSS (mm)	0.004	0.004	0.005	0.005	0.004	0.004	0.005	0.003	0.004	0.004
SED (T/HA)	0.005	0.006	0.005	0.006	0.004	0.004	0.005	0.003	0.003	0.004

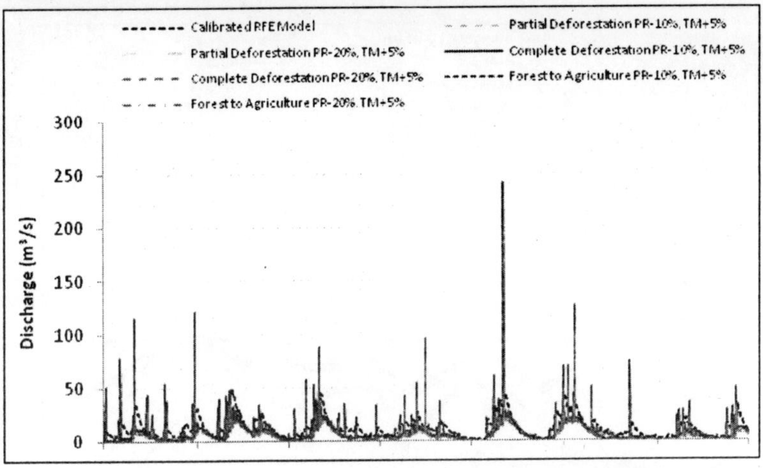

Figure 8.7: Daily Discharge for Land Use-Climate

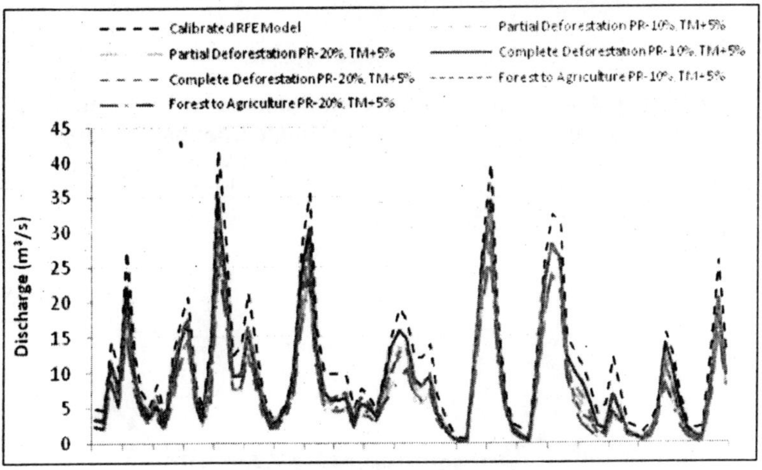

Figure 8.8: Monthly Discharge for Land Use-Climate Change Scenarios

Annual Average Percent Changes in Water Balance Components for Climate and Land Use-Climate Change Scenarios

The Figures 8.9 and 8.10 show the percent changes in the annual average water balance components for the climate and land use-climate change scenarios. From Figure 8.9, it is evident that sediment yield is the most responsive followed by revap, surface runoff and transmission losses. For the land use-climate change

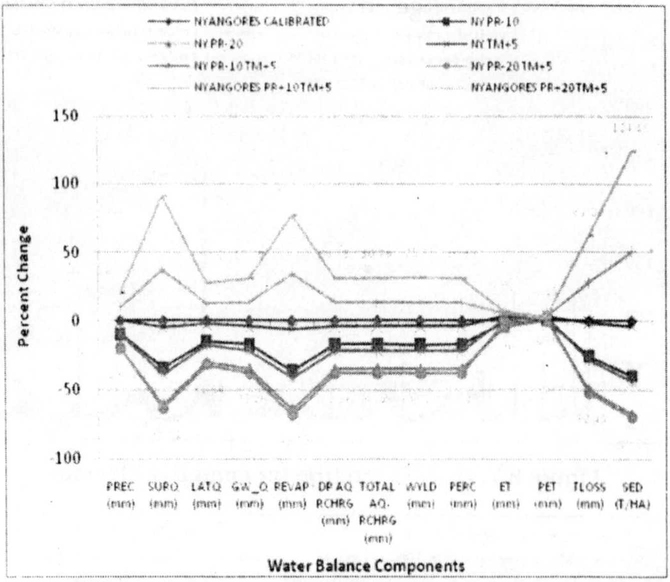

Figure 8.9: Annual Average Per cent Changes for Sal Watershed Water Balance Components for Climate Change Scenarios

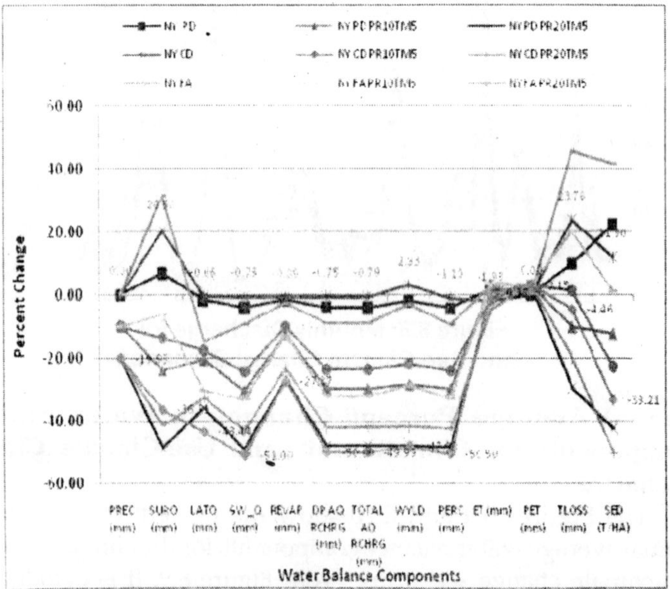

Figure 8.10: Per cent Changes for Water Balance Components in Sal Watershed Land Use-Climate Change Scenarios

scenarios, the percent changes in water balance components in Figure 8.10 below display the variation in the water balance components across different land uses. Details of these changes are shown in detail in Tables 8.11 and 8.12, the different land use scenarios affect the water balance components differently and where these differences are most pronounced are in the surface runoff and groundwater recharge.

All the water balance components vary linearly to precipitation, which can also be observed from the plots that are almost identical to one another, especially those that are of corresponding reduction/ increase in precipitation. In the land use-climate change combined scenarios, there is a reduced amount of groundwater recharge, surface runoff, and total water yield to the stream meaning the water balance will be significantly affected by the reduction of precipitation, increased temperature and altered land cover.

The climate change scenarios revealed that the variation of precipitation has the greatest impact on the amount of discharge, sediment yield, surface runoff (a reduction of 20 per cent in precipitation will reduce the surface runoff by half its amount in Sal River watersheds) and generally to the water balance components in the watersheds. The ratio of the water balance components to precipitation reduces drastically with the reduction of precipitation. This is expected as precipitation is the main driving force of the hydrological cycle and any change in the amount will be directly reflected in the flow of the Sal River.

Temperature increase affects the discharge less directly than decrease in precipitation but nonetheless has an impact by increasing evapotranspiration and plant production, which will ultimately affect land cover in the long-term. More realistic climate change simulations that combined precipitation reduction and temperature increase had the most effect on the discharge and water balance components with a reduction in river discharge in both wet and dry seasons, reduction in total water yield, groundwater discharge, surface runoff and groundwater contribution to the river channel.

Conclusions

Based on the results obtained in this study, the visual interpretation is a suitable methodology to classify imagery at a high and produce an accurate map of a highly variable area using far less time and effort than conventional algorithms. The resulting map obtained from this land cover classification was of a high accuracy (85 per cent) and was suitable for use as an input into SWAT especially for the simulation of effects of land use change in a spatially explicit hydrological model.

The model evaluation results (Tables 8.4 and 8.5) suggest that the calibration process may have not adequately captured the variations in the different hydrological years (periods) in Sal Watershed (Chamba) the Rain gauge and RFE models which may be due to the fact that the time series data was not long enough to achieve this. In the case of the Rain gauge models compared to the RFE models, the statistics and hydrographs show the rainfall values from the Rain gauge data were not well representative of the actual rainfall that was received in the watersheds under study. Lack of a dense rain gauge station network within the study area that was unable to capture the different rainfall amounts and account for the spatial variability of the rainfall received is the most likely cause of this result. Rainfall is the main driving force of the hydrological cycle and when the rainfall for large watersheds such as the Sal watersheds cannot be accurately accounted for this presents a problem in the simulation process and when calibrating the model because this necessitates the rigorous adjustment of parameters which is not only a time consuming process but also may result in parameter values that may give a good simulation result but are hydrologically unrealistic for the watershed.

However, it can inferred that the set-up and calibration of a semi-distributed hydrological model such as SWAT in a large watershed with variable land cover, soils and topography is a feasible task and will yield satisfactory results given reliable data and proper attention to manual or automatic calibration.

The model simulations showed that the Sal River flow would be significantly affected in the face of the climate and land use change scenarios posing difficulties in adaptation to the altered flow regimes of the Sal rivers.

The different water balance components were affected regardless of the type and amount of change that was undergone thus affecting the magnitude and timing of the flow. It is therefore prudent to work towards establishing and maintaining adequate minimum flows that would mitigate the effects of reduced base flows and put in place measures to maintain adequate sustained river flows to the benefit of the stakeholders of the Sal River Watershed such as proper land and water management practices.

References

Abbaspour, K.C., Yang, J., Maximov, I., Siber, R., Bogner, K., Mieleitner, J., Zobrist, J. and Srinivasan, R., 2006. Modelling hydrology and water quality in the pre-alpine/alpine thur watershed using SWAT. *J. Hydrol.*, 333: 413–430.

Anderson, R.J., Hardy, E.E., Roach, J.T. and Witmer, R.E., 1976. A land use and land cover classification system for use with remote sensor data. *United States Geological Survey*, United States Government Printing Office, Washington D.C.

Di Luzio, M., Srinivasan, R. and Arnold, J.G., 2002. Integration of Watershed Tools and the SWAT Model into Basins. *J. Am. Water Resour. Assoc.*, 38(4): 1127–1141.

Ficklin, D.L., Luo, Y., Luedeling, E. and Zhang, M., 2009. Climate change sensitivity assessment of a highly agricultural watershed using SWAT. *J. Hydrol.*, 374: 16–29.

IPCC–Intergovernmental Panel on Climate Change, 2007. *Climate Change Synthesis Report.* Cambridge Press, Cambridge.

Moriasi, D.N., Arnold, J.G., Van Liew, M.W., Bingner, R.L., Harmel, R.D. and Veith, T.L., 2007. Model evaluation guidelines for systematic quantification of accuracy in watershed 5 simulations. *Transactions of the ASABE*, 50(3): 885–900.

Neitsch, S.L., Arnold, J.G., Kiniry, J.R. and Williams, J.R., 2005. Soil and water assessment tool. *Theoretical Documentation: Version 2005, Temple, TX.* USDA Agricultural Research Service and Texas A and M Blackland Research Center.

Ogutu, O.J., Piepho, H.P., Dublin, H.T., Bhola, N. and Reid, R.S., 2007. El Nino-Southern oscillation, rainfall, temperature and normalized difference vegetation index fluctuations in the Mara-Serengeti ecosystem. *Afr. J. Ecol.*, 46: 132–143.

Pareta, K., 2003. Morphometric analysis of Dhasan river basin, India. *Uttar Bharat Bhoogol Patrika*, Gorakhpur.

Pareta, K., 2004. Hydro-geomorphology of Sagar district, M.P.: A study through remote sensing technique at Dr. Hari Singh Gour University, Sagar (M.P.) in XIX M. P. Young Scientist Congress.

Pareta, K., 2004. Study of soil conservation and water resources management of Bina river basin, M.P. *National Geographical Research Conference (NGRC).*

Pareta, K., 2005. Rainfall-runoff modelling, soil erosion modelling, water balance calculation, and morphometric analysis of Molali Watershed, Sagar, Madhya Pradesh using GIS and remote sensing techniques. *Presented at the 25th International Cartographic Congress, INCA.*

Pareta, K., Chaudhery, K., Hansen, L. B., and Sorensen, M.K., 2008. Land use and land capability report in emergency tsunami reconstruction project: Preparation of an integrated coastal zone management plan and coastal vulnerability maps of Tamil Nadu. *Unpublished Project Report,* Department of Environment, Government of Tamil Nadu

Pareta, K. and Upasana, Koshta, 2009. Soil erosion modeling using remote sensing and GIS: A case study of Mohand watershed, Haridwar. *Madhya Bharti Journal,* Dr. Hari Singh Gour University, Sagar, M.P., 55: 57–70.

Pareta, K., 2010. Advance remote sensing and GIS technologies for hydrological study and groundwater exploration in Bhansar and Budhil watershed, Himachal Pradesh. Applications of Geoinformatics for National Development, Department of Applied Geology (Center of Advanced study in Geology), Dr. Hari Singh Gour Vishwavidyalaya (Central University), Sagar, M.P.

Refsgaard, J.C., 1997. Parameterization, calibration, and validation of distributed hydrological models. *J. Hydrol.,* 198(1): 69–97.

Strahler, A.N., 1952. Dynamic basis of geomorphology. *Bull. Geol. Soc. Am.,* 63: 923–938.

Thornbury, W.D., 1990. *Principle of Geomorphology,* 2nd Edn. Wiley Eastern Limited, New Delhi.

Wentworth, C.K., 1930. A simplified method of determining the average slope of land surfaces. *Am. J. Sci.,* 21: 184–194.

Chapter 9

Direct Negative Impacts of Micro-Hydel Projects on Poor Farmers: A Farmers' Perspective

☆ *Ramesh Ganeriwal*

ABSTRACT

Unknown to most, Micro-Hydel projects in the hilly states like Himachal Pradesh are doing enormous harm to farmers in terms of either completely taking away the irrigation water or in some instances, changing the ever-flowing water schedules to that which suit the upstream power producers, thereby causing a great deal of disruption to the farming operations. This paper describes the plight of farmers, who are too poor and unorganized without the wherewithal to obtain redressal for the gross injustice that these relatively recent water diversions are inflicting on the farmers who have enjoyed the rights to irrigation waters for generations. The nodal agency designated by the Himachal Pradesh Government, Himurja is less than forthcoming in providing requested information by the aggrieved parties to be able to monitor the compliance of stringent environmental and other norms imposed on the project builders and the operators.

Even the "run-of-the-river" projects, which every one imagines to be innocuous to ecology and human population, the water diversion causes dry or much lowered water level stretches of rivers/nallah's that take away traditional water sources from the age-old water channels [called 'kuhls' in the local parlance], thus causing severe hardships on poor and marginal farmers. The 15 per cent water "rule" that is supposed to be left in the river by the power producers is arbitrary and serious debate is the urgent need of the hour so that appropriate amount of water is provided in accordance with the irrigation water needs of the farmers – otherwise, the farmers are literally left "high and dry" without their day in the court!

Keywords: *Micro-hydel projects, Farmer, Kuhal governing laws.*

Introduction

The new onslaught of the mini and micro-hydel projects, so enthusiastically promoted by the Government has completely ignored the ancient right of the farmers to the irrigation water and is causing untold harm to small and poor farmers of the hilly states.

This paper presents two examples where the irrigation water of a vast number of farmers has been disrupted by a State Electricity Board with no due-process of obtaining "No Objection Certificate" (NoC) private mini-hydel project is presented where the owner of the project deviously obtained NoC from a different Panchayat whereas the affected farmers had been left out of this mandatory process and the farmers now face a peril to their standing crops because no water is left in their "kuhl" while causing harm to fish and other ecological aspects of the area.

Finally, the so-called 15 per cent "rule", which mandates the power project operators to leave for the use of the down-stream users is questioned with an appeal to the Government authorities to examine the efficacy of this rule and the mechanism whereby the compliance of this rule can be monitored while there is debate and settlement of the water rights that have been enjoyed by the farmers from time immemorial.

Water Diversion Hardships imposed on the Farmers

Two cases of the subject matter are presented in this section. First, in sub-section 2.1 is the case of a State Electricity owned mini-hydel project where a large area of Kangra District farmers have been affected with the disruption in the water schedule of their traditional irrigation water "kuhls'. Section 2.2 presents the case of a recently commissioned private micro-hydel project where the affected farmers have been left with no water what so ever and as a consequence, these farmers face the prospects of completely losing their standing wheat crop while the process of redressal for this great injustice is initiated in the appropriate forums.

Irrigation Water Scheduling Disruption

The "Kholi Khud" mini-hydel project in Kangra District is an interesting case where the State Electricity Board has taken no cognizance of the hardships and disruption that would be caused to approximately 2500 farmers, covering an area of over 2500 hectares of 12 Panchayats from 4 "kuhls" serving these farmers. The project was ill-conceived in that there is not enough flowing water that will allow the project to generate power continuously. So, after the fact, the project now accumulates water on a regular basis, thereby disrupting the irrigation water supply and when there is sufficient water accumulated to generate power, the water is released in the "kuhls". This causes severe disruption to the water needs of the farmers, where some times, the sowing timings of the crucial crops are impacted and most other times, farmers end up watering their fields at nights and so on. It is interesting that the Electricity Department has taken no cognizance of the concerns of the farmers, either before undertaking the project (in terms of securing NoC's from the concerned Panchayats) or in the current operations of the project.

All Irrigation Water Diverted by a Private Micro-Hydel Project

One of the more egregious examples of having irrigation water completely taken away from the farmers is presented in this subsection. This is a Micro-Hydel project of approximately 1.5 MW capacity built by a private operator that has been recently commissioned. The concerned farmers have made representations to the District Authorities during the construction phase of this project but there was no action taken on their applications. The evasive response from Himurja on a request for information on this project will be illustrated in the next section of the paper but the net result today is that when the project was commissioned about two months back, absolutely water was left in the "kuhl" that provides irrigation water to these farmers. Interestingly, the project builder did go through the motions of taking NoC from their Panchayat but it turns out that the affected farmers belong to another neighbouring Panchayat and hence, they were never consulted in this process. Presently, these farmers face the grim prospects of losing their standing wheat and other crops and have made an urgent appeal to the DC, Kangra for the immediate restoration of their water while the matter is taken up in appropriate forums. The following pictures show the river when the river was in full flow followed by a picture of dried "kuhl".

Figure 9.1: Showing Availability of Water in Gaj Khad where Power Project is Proposed

Last, but not the least, there is ecological damage caused by the sudden lowering of the river water level, that seems to fallen out of the purview of such authorities as the Fisheries Department that are responsible for maintaining the health of wild life. The following picture shows dead fish that were left in the aftermath of the water diversion from this particular location.

Figure 9.2: Showing Water Diversion in Kuhls by Farmers to Irrigate their Land in Gaj Khad

Evasion of Statutory Information to be Provided under the RTI Act

This subsection illustrates an example of how a State Government agency, namely Himurja, stonewalls and evades request for legitimate information under RTI act. Exhibit I at the end of the

Figure 9.3: Showing Dead Fishes because of Sudden Stoppage of Water in the Downstream of Power Project in Gaj Khad

paper shows the original request by a concerned social activist for the above project during its construction phase. Exhibit II shows the first response of Himurja to this request wherein the agency essentially denied this request under the guise of this information being "3rd Party information" requiring the consent of the company. Exhibit II shows the stern follow up letter by the applicant saying that most of the information requested is public in nature because it deals with diversion of public goods. Finally, Exhibit III shows the final response where Himurja partially (and reluctantly) provided partial information while awaiting the consent of the project owner.

Such evasion clearly puts a question mark on Government's sincerity in allowing independent as well as the aggrieved parties to be able to monitor the compliance of the norms that are imposed on the project builders to safeguard environmental and other sensitive issues.

Efficacy of the 15 per cent "Rule"

In the official circles concerned with issues of permitting water diversion for hydel projects, there appears to be a belief that there is a "15 per cent rule" whereby the project builder is supposed to leave 15 per cent of the water volume in the river for downstream users and can lift as much as 85 per cent of the water.

The efficacy of this "rule" needs serious debate and consideration in light of the fact that most of the times, the downstream users are farmers who have enjoyed the rights to irrigation waters for generations. In the interest of fairness and justice, it is absolutely important a serious debate, involving all stake-holders takes place where all aspects of this conflict of water use is brought to the forth. On the face of it, it is clear that 15 per cent water is not adequate for the farmers to maintain their traditional cropping patterns. Further, it must be investigated what the exact definition of this water quantum is – for example, if it is taken to be the average of the water flow over a year, then, it will absolutely not serve the farmers because the irrigation requirements are concentrated over certain time periods in the crop cycles whereas there is lot of overflow of water during rainy season for example. It would seem fair that the consideration of this matter begins with the actual water requirement of the farmer and then guidelines are developed to provide the "surplus" water to the power project. In these deliberations, it may evolve that there are potential water conservation techniques that will provide more water for power generation but the burden of financing and implementing those measures must not fall on the farmer, but rather, this investment should be Government's and Power Generator's responsibility.

Conclusions

Even mini and micro-hydel projects, which were considered to be relatively harmless to the humans and ecology, are causing great deal of hardship and injustice to the farmers. The rules and regulations framed for diverting water for power generation are flawed and are not implemented and monitored properly. It is the urgent need of the hour for urgent actions at all levels of the Government so that the poor and marginal farmers are not driven into further hardship and displacement.

Chapter 10

Allahrakhi: A Case Study of Community Managed Canal Irrigation System

☆ *Sachin Kumar and Shailendra Bist*

ABSTRACT

Common property resources as an institutional arrangement for managing resources are strikingly different from free and open access with which they are increasingly being confused. The case examines the issues and problems related to the community managed canal irrigation system in Markham Grant in Dehradun, Uttarakhand.

The indigenous institutional arrangements like these are coming under pressure from the forces of increasing commercialization of resources which were meant for subsistence economies. They are increasing under threat of being converted into private property either through privatization policies of the state or covert acquisition by the vested interest.

The case illustrates the use of design principles of managing long enduring, self-organized irrigation institutions.

Introduction

Allahrakhi canal he felt the anxiety of managing the community owned canal irrigation system. He thought of the increasing pressures from the fellow irrigators, populist. The sowing season for paddy had just started and as Jal Prabhandak[1] walked besides the policies of the government and the unreliable monsoon. It was only in the last week during the annual maintenance of the canal that they had to call an Emergency meeting of the Jal Samiti[2]. One of the household had sent a young girl and not the ablest male member, as is the practice of the community, for the maintenance work. This was resented by other households as a fit case of free-riding. The girl's father quoted the example of a small land holder who did not even contribute even a single day of labour and yet no action was taken against them. The practice in the Chandmari branch of the irrigation system, a household owning less than two bighas[3] of land has to contribute one labour-day and if the holding is more than five bighas, the households have to contribute labour for the entire duration of the maintenance work. All household users have to share nominal cost of maintenance in proportion to their land holdings. After this dispute, the Jal Pradhan seriously thought of codifying the interpretation so as to prevent a rather subjective interpretation of the rules.

Lying at the foothills of the middle Himalayas in the Uttarakhand, Markham Grant (Figure 10.1) is a predominantly agricultural community. Demographically, it is inhabited by Hindus towards the north, Sikhs in Centre and Muslims towards South in the irrigated area of Allahrakhi. There is absence of local Garhwali population in the area mainly because of the fact that the grant of Markham was made by the British in 1838 to ex-army men or the outsider for their contribution to the British Raj. The Allahrakhi had been constructed by them after the grant of the land and had been irrigating the lands since then.

---

1 Literally, Water manager. He is head of the body which manages the community owned canal.

2 Literally, Water Committee. It consists of users and meets regularly to decide the distribution of water and in emergency to settle disputes, if any

3 1 **acre** = 2.5 **Bighas** = 1600 m2, 1 hectares = 6.25 **Bighas**

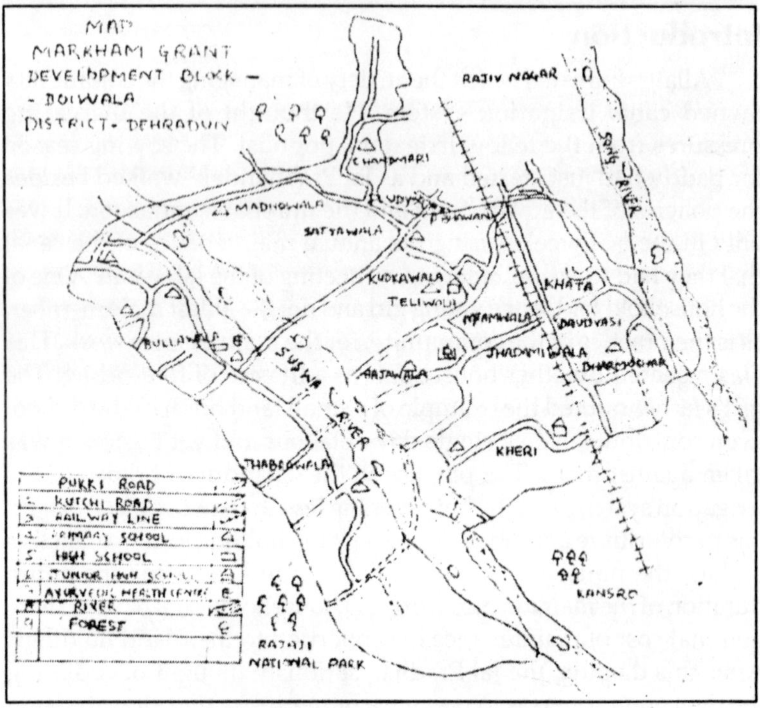

Figure 10.1: Map of Markham Grant and Adjoining Areas

Located in the vicinity of Dehradun city, Markham Grant has not been left untouched by the increasing commercialization of the erstwhile subsistence economy. Being close to the national highway has also brought increasing mechanization and commercialization of the agriculture in Markham Grant. With mechanization of agriculture and lucrative prices of sugarcane (in fact subsidized by the government) his fellow villagers have become greedy and area under sugarcane cultivation has drastically increased.

With increasing pressure on land due to increasing population, the water resources are also strained. The disputes and altercations on the issue of priority as well as quantity of irrigation water are increasing. Their emergency conflict resolution meetings are becoming less successful in arriving at results. Although users have threatened to go to the civil courts for resolution of the water sharing disputes but he has been spared the embarrassment till now. This could be troublesome as the management committee of the Allahrakhi canal is not recognized by the government.

The Jal Pradhan at times is financially helpless as all grants/ aid/schemes of the government for improvement of irrigation facilities for his area are routed through the Panchayat Samiti or in fact through the Gram Pradhan[4] himself. This creates resentment among different religious and caste groups as the subsequent allocation of the grants takes political and communal colours.

Funds are also available to strengthen the agricultural infrastructure by making the canals in the area, including Allahrakhi, fully lined. The non-lined nature of the Allahrakhi Canal leads to loss of efficiency in carrying capacity of the water as the efficiency of unlined canals is only 60 per cent as compared to 70-75 per cent of fully lined canals[5].

Jal Pradhan understands fully well that an efficient management of common property resources (CPRs)[6] like water requires clearly defined boundaries. Therefore, the rights to use the water are linked to land holdings and the water supplied through the canal is directly proportional to the land holdings. For example, Chandmari village which is at the head of a sub branch of the canal follows a simple water allocation rule- one hour of sub canal water for ten bighas of land. The organization of the whole system has been kept simple. A work team of four or five households form the first tier. The second tier may be of all the users of a particular branch of an irrigation system. The third tier is the collection of all the households of one headwork. The small teams prevent free riding andlarge enterprises provide economies of scale and aggregation of capital.

The Jal Prabhandak has carried forward the intergenerational collective choice arrangements for the use of irrigation water. In fact, the very spirit of community managed CPRs asks for inclusion of all

---

4 Literally, Village Headman. He is elected through popular vote of the unit once in five years.

5 Irrigation Efficiencies, www.and.nic.in/page7.pdf

6 CPRs as the common property resources are commonly referred to as are the resources which are accessible to whole community to which no individual has exclusive rights.

the stakeholders in the decision making body. The absence of a statuary authority makes it imperative for the users to formulate effective rules which are in consonance with specific characteristics of their settings. In the Allahrakhi canal management, all decisions are taken by consensus and all monitoring is done by the users themselves. They had explored the idea of an external/non-user agent for monitoring purposes but rejected the idea by consensus. They felt that such an official would require some compensation to catch and/or report a malpractice by any of the users. Also Jalprabhandak takes care of monitoring the rotation of irrigation water among the users of particular branch. For the management of actual usage, it is almost self-regulatory and violations of the operational rules in community managed canal irrigation system at Markham Grant are met with graduated sanctions. In the case of Markham grant, for example, in the water rotation system, the irrigator who nears the end of a rotation would like to extend the time of his turn (and thus extend the amount of water obtained) but the next irrigator in the rotation system waits for him to finish and would like to start early. Thus the presence of first irrigator deters the second from starting early and that of second form ending late for the first.

Increasing Government involvement has brought unheard of situations which their forefathers had neither faced nor had thought of them. Recently landless villagers have started asking for a share in the water rights of the canals as is the case with some of the crafted institutions in the country which is opposed by traditional right holders. The land holders view this as a ploy to create dissent in the community. They feel that these landless villagers already use the canal water for non- irrigation purposes like cattle rearing, washing, cleaning etc. without contributing to the maintenance etc.. Some of the smaller landholders and landless villagers with covert support of Gram Pradhan have started clamouring for taking the support from the government.

More trouble is brewing in the neighbourhood for the community and the canal. The pressure created by the civil rights group has forced the government to develop another site for rehabilitation of the dam- displaced people. The new site would be located close to the Markham Grant as the land is fertile and the perennial source of water is available in the form of Song River. There is talk of getting another government canal from the river and it could be sourced closer to the origin of the rivulet from where Allahrakhi also draws water.

He remembered meeting one of the government dignitaries, few years back, who wanted to replicate this model across the length and breadth of the country. He had to tell him bluntly that we are a small and cohesive community; it may not be possible to replicate the same process in every village of the country. In the changing scenario, he is not sure how long would he be able to call themselves as a 'small and cohesive community' when the pressures of commerce and politics are tearing at the fabric of their community.

The monsoon has been pretty predictable and generous this year but Jal Pradhan wondered how long can they manage the growing needs and aspirations of the users of the Allahrakhi? Would he be forced to see the privatization or take-over of the Allahrakhi by the government?

Chapter 11

Climate Change and Water Resources: Issues and Options

☆ *Vivek Trivedi*

Unplanned development, concrete surfacing, reclaiming of land and other human induced action has impinged water bodies, river course, catchments. Mainly human factors are responsible for breaking the silence of nature; troubling flora and fauna, littering non-biodegradable waste in the pristine glaciers, throwing untreated sewage, industrial waste chemical in the sacred perennial river originating from great Himalayas.

Non- serious approaches, untenable policies, skimpy implementation due to bureaucratic inertia and lack of political commitment have certainly hampered the success of pilot project initiated and in replicating the successful models for the conservation of natural resources *viz.* frozen glaciers, dense forests, wild life (flora-fauna) etc.

Gradually with advent in high pace of uncontrolled industrialization and other factors have changed Himalayan climate a lot. The mountain ecosystems harbor a wide range of natural resources and are particularly sensitive to change. Regional changes in climate have already affected a number of the physical and biological systems in the mountains. Analysis of temperature trends in the Himalayas and its vicinity shows that temperature increases are greater in the uplands than the lowlands.

From the studies covering over 30 years of records, average air temperatures were found to be 0.7 to 2.4°C higher than that in the 1980s, as against the global average of 0.5°C; the Himachal Pradesh trend indicates an increase of 0.06°C per year. An analysis of rainfall data over the period 1976 to 2006 show increasing trends of rainfall in Lahaul, Spiti, Chamba and Kangra but decreasing trends in Solan and Kinnaur. Other districts showed no signicant trends. Crops are showing shorter periods of owering and maturity due to change in the climatic patterns.

Observational records and climate projections provide abundant evidence that freshwater resources are vulnerable and have the potential to be strongly impacted by climate change, with wide-ranging consequences for human societies and ecosystems. Observed warming over several decades has been linked to changes in the large-scale hydrological cycle such as: increasing atmospheric water vapor content; changing precipitation patterns, intensity and extremes; reduced snow cover and widespread melting of ice; and changes in soil moisture and runoff. Increased precipitation intensity and variability are projected to increase the risks of flooding and drought in many areas. Higher water temperatures and changes in extremes, including floods and droughts, are projected to affect water quality and exacerbate many forms of water pollution – from sediments, nutrients, dissolved organic carbon, pathogens, pesticides and salt, as well as thermal pollution, with possible negative impacts on ecosystems, human health, and water system reliability and operating cost. Climate change affects the function and operation of existing water infrastructure – including hydropower, structural flood defenses, drainage and irrigation systems – as well as water management practices. Current water management practices may not be robust enough to cope with the impacts of climate change. Mitigation measures can reduce the magnitude of impacts of global warming on water resources, in turn reducing adaptation needs. The paper will be detailing the state of water resources in the Himalayan context with an open option to mitigate such situation by taking into account the best model practiced *viz.* Better water management, Recycling of water, Efforts for water wastages, Inclusion of Water Harvesting in building bye laws, Use of tertiary treatment of water as in Chandigarh, Progressive recovery of water charges, PPP in water supply and distribution, check dams, Deep wells to store water surface runoff, Monitoring of Water Supply (Supervisory Control and Data Acquisition) System in MC Chandigarh etc.

Chapter 12

Sustainable Development in Context of Climate Change

☆ *P.K. Gupta*

The concept of sustainable development got prominence in the decade of 80;s and it refers to the process of socio-economic development through the judicious utilization of the natural resources, keeping in view the needs of future generations. Socio-economic development is considered to be very important for the society. The element of 'sustainability' comes from the initiatives of environmental protectionist. This process keeps its eye on the earth's carrying capacity. It has been realized that the consumption of natural resources is higher than that of their regeneration. So the concept of sustainable development emerged out of the fears of depleting natural resources. Nature has bestowed human's societies and economies with a complex life support system air, water, food and suitable climate for survival. Nature has also supported and maintained life on earth since time immemorial and should continue to do so in the future. Sustainable development is a value based concept which appeals to the universal themes of mutual co-existence and respect for others.

Sustainable development is continuously evolving process bringing together cultural, social, economic, environmental and political concerns. Sustainable development has three fundamental principles of inter-generational equity and justice, intra-generational equity and justice and the respect for the carrying capacity of the land which means the ability of the earth to maintain human beings sustainably and indefinitely. The basic human needs are water, air, food, clothing and shelter that are supported by the earth.

The countries of the world are inter-connected through the global environment with the activity of one country affecting the others. The industrialized countries have just 20 per cent of the World population but consume about 80 per cent of the World's resources. They contribute a major share of the global Environmental degradation process both through exploitation of natural resources and pollution.

Climate change is a serious threat faced by the International community striving towards sustainable development. It has implication not only for health and well-being of the earth's eco-system but also for the economic enterprises and social livelihood. The global warming has become great challenge to the mankind in the form of increasing drought and desertification, crop failure, Coastal flooding, melting of ice caps, natural disaster, infectious diseases etc.

Water is one of the most crucial natural resources. Life on the earth depends upon water. There are two types of water resources *viz.* surface water and ground water. Surface water consist of rivers, streams, lakes etc whereas, ground water consists of wells and springs. The sustainable use of water requires that the hydro logical cycle not to be disrupted. There is evidence that large scale deforestation causes serious disruption in Rainfall patterns. The removal of vegetative cover in catchment areas causes, floods and droughts. Man's activities have disrupted the hydro-logical cycle and contaminated both the surface and ground water resources. Acid rain is the cause of atmospheric pollution. Polluted water also poses threat to human health.

Aqua care is the need of the hour. Aqua care means protection of water resources, involving prevention of pollution as well as cleaning of surface resources such as tanks and rivers and ground water, similarly keeping the sea water un-polluted and clean so that the life of aquatic creature is not in danger and prevention of the spread of water borne diseases by ensuring safe drinking water for all. It is civic duty of every citizen to do everything to achieve the objective of aqua care spelt out above. Aqua care constitutes an important subject in the curriculum of 'EDUCARE' so that the children become aware of the importance of aqua care and they too contribute towards the achievement of the goals of aqua care.

Therefore, the concept of sustainable development needs to be incorporated as an essential part in each and every development policy framework of national governments and international agencies.

I take this opportunity to thank the (UGC) university grant commission for extending financial support for organizing this conference. I am also thankful to the organizer and co-organizer *viz.* P.G.College Chamba and Mountain Forum Himalayas Shimla for organizing national conference on sustainable use of water resources in context of climate change-A shared responsibility. I end my key note address with the theme that we should, 'Think globally and act locally'.

Chapter 13

Community Management of Groundwater: Basis for Piots

☆ *Amit Upmanyu*

ABSTRACT

Groundwater, although an established „common resource, is treated almost as a private resource because of its decentralized and easy access at the individual level, particularly in India. The definition of groundwater as a common pool is perhaps the first step in establishing it as such. While at the definition level it is easy to talk about resources like forest, rivers, common land etc. because there common ownership or common use is visible and established. In case of groundwater whose existence, occurrence and mobility give it a halo of „uncertainty its abstraction and application almost attribute an individually or privately owned status. With this uncertainty, and the need to push for community management of ground water, we need to generate sufficient knowledge about aquifers, aquifer boundaries, recharge and evolve practical mechanisms of sustainable abstraction which are equitable, sustainable and most importantly, acceptable to the community. Creating knowledge, developing the mechanisms of implementation and piloting such processes at numerous locations in different situations is required to develop a strong case for the community management of groundwater, especially in the Indian context. ACWADAM, along with various partner NGOs, in different parts of the country, is trying to run trials under the necessary guidelines for community management of groundwater as a CPR.

This paper broadly describes the current status of groundwater in India. It looks groundwater as a common pool resource, explains the need of community management of this resource and also suggest protocols and management strategy for community based groundwater management.

Keywords: *Groundwater, Community management, Sustainable utilisation.*

Background

Water is an extremely important component in the entire earth system and it is vital for the basic functioning of the entire system. It moves from one sphere to the other in a cyclic form which is called hydrologic cycle. During it journey, water takes different forms in different spheres and is utilized by humans, animals, vegetation and other living beings. It also provides a balance to each of these spheres of the earth system. We can divide the entire available water into two main categories a) saline water and b) fresh water. Around 94 per cent water on the earth is saline water and remaining 6 per cent is fresh water. Fresh water can be further divided into three categories a) groundwater which is 71.23 per cent, b) ice or glaciers which is 28.49 per cent and c) surface water which contributes 0.28 per cent to the total fresh water available on the earth (Freeze, R.A. and Cherry, J.A. 1979).

In India 38.5 per cent of the total available fresh water comes from groundwater and it meets 55 per cent of irrigation, 85 per cent of rural and 50 per cent urban and industrial water requirements (Planning Commission, Government of India, 2007), which shows the huge dependency on groundwater in our country. This need is fulfilled by thousands of wells, bore wells and springs in the country. Rapid growth in population has increased the water demand which has resulted in a high level of abstraction of groundwater and depletion of groundwater systems *i.e.* aquifers, in the country. Groundwater abstraction saw a leap in utilization during the green revolution period of 1960-65. Since then, the groundwater extraction has been increasing and it has resulted in many of these prime grain growing areas of the country to face

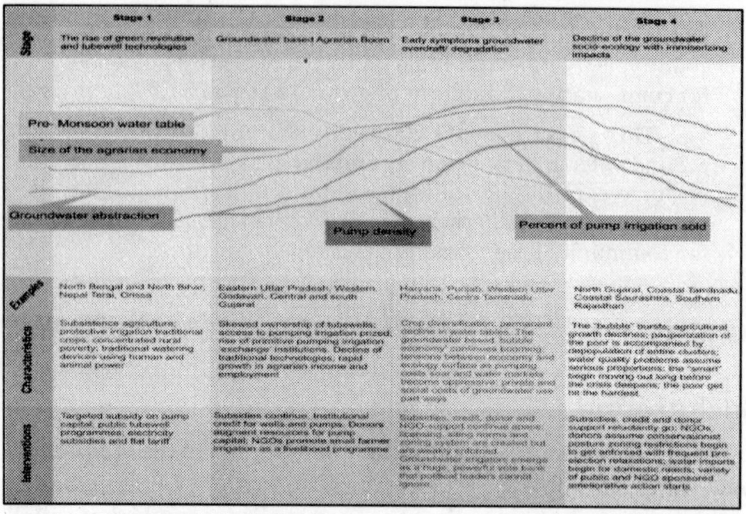

Figure 13.1: Tushar Shah's Diagram Showing the Four Phases of Groundwater Development in South East Asia

groundwater problems of quantity and quality. These areas have been declared to be over-exploited by Central Ground Water Board (CGWB). In his book, Taming the Anarchy, Tushar Shah has described the stages of groundwater development in India over the period of last 50 years. In these 50 years, the number of wells bore wells and the amount of groundwater abstraction has gone up enormously and consequently the groundwater levels have gone down significantly.

Pattern of Groundwater Use in India

India has the largest number of private wells and bore wells (27 million) (Taming the Anarchy) in the world and the largest groundwater irrigated area. Every third farmer in India owns a private irrigation well/bore well. These figures themselves reflect groundwater dependence and pattern of groundwater use in the country. Farmers are free to dig a well on their field and abstract as much water as they want. The Easement Act, 1882 gives them the right over groundwater that is found under their land.

There is no control over well or deep bore well digging for that matter and farmers have all the freedom for this, except in areas which have been declared as dark zones by CGWB (farmers are not allowed to drill bore wells or get 3 phase electricity connections in such areas). Groundwater is considered as the most reliable, easily accessible and decentralized source of irrigation and owning a well or bore well is considered as a status symbol. Number of wells/bore wells increase with the increase in landholding divisions as every farmer wants to own his individual well/bore well. Groundwater is considered and treated as private property and there are well established practices of groundwater marketing in many parts of rural India. This sense of private ownership makes groundwater different from all other common pool resources like forest, rivers, common lands etc.

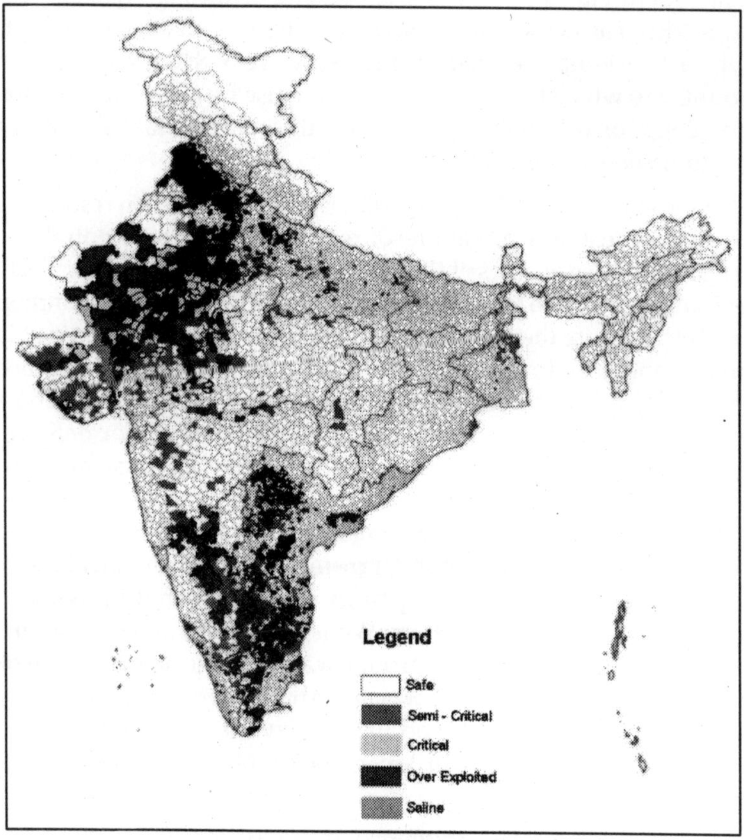

Figure 13.2: CGWB Map Showing the Status of Groundwater Exploitation in India

Groundwater Common Pool Resource

Common Pool Resources (CPRs) are natural or human-made resources where one person's use subtracts from another's use and where it is often necessary, but difficult and costly, to exclude other users outside the group from using the resource (Governing the commons). CPRs are those resources that cut across the administrative and other such boundaries and are common to many people and communities. Two main characteristics of CPRs are that their spread is not defined by administrative or political boundaries and they are used by a large group. For example, a river may flow through many districts of a state, many states of a country or sometimes through two or more countries. Throughout its path, river water is used by many people who are residing near that river irrespective of the district, state or country to which they belong. Common use by a large population and common ownership of this natural resource establishes it as a common pool resource.

Groundwater, although considered to be a common resource is treated almost as a private resource because of its decentralized nature and ease of accessibility at the individual level, particularly in India. It is easy to talk about resources like forest, rivers, common land etc. because their common ownership or common use is visible and established. In case of groundwater, its existence, occurrence and mobility is uncertain but its abstraction and application almost attribute an individually or privately owned status. For defining groundwater as a common pool resource we first need to understand the occurrence and basic characteristics of groundwater. Groundwater is found in aquifers, which is a water bearing rock. Aquifers store and transmit sufficient amount of water to the wells, bore wells and springs. Water is stored in the pore spaces and fractures in aquifers. Wells, bore wells and springs are the mechanism to access the groundwater but the actual source of groundwater is the aquifer in which the respective well, bore well or spring exists. Groundwater occurrence, available quantity and its transmission to the wells will depend on the physical characteristics of the aquifers.

Figure 13.3 contains aquifers in the sandy layers which extend across several villages. Figure 13.4 contains individual villages lying over small aquifers.

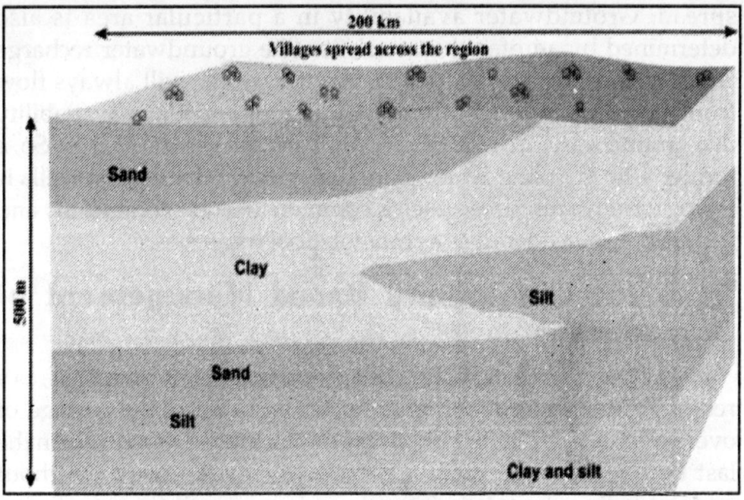

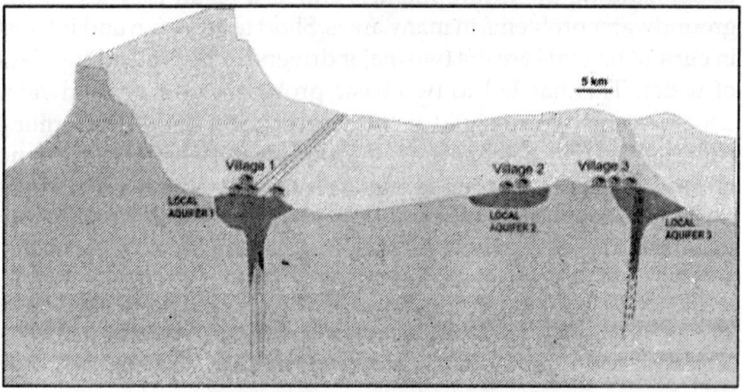

Figures 13.3 and 13.4: Nature of Spatial Variation of Aquifers

An aquifer is like an underground watershed and its boundary can sometimes coincide with the surface watershed while at other times it can extend through many watersheds, villages or blocks irrespective of administrative or political boundaries. Similarly, it does not coincide with the private land holdings but its water can be accessed at individual farm level. Wells in the same aquifer are connected to each other and water extraction from one well affects the water level in the nearby wells. This zone is called radius of influence, this influence can be local or sometimes very wide spread. Groundwater availability in a particular area is also determined by aquifer characteristics like groundwater recharge and discharge zones. Within the aquifer, water will always flow from the recharge zone to the discharge zone and water availability in a groundwater discharge zone will always remain higher. So, a farmer who is taking water from his individually owned wells is taking it from a resource which is common to larger community and it should be considered as a common pool resource.

Need for Community Based Management of Groundwater

At present groundwater abstraction is at an all-time high, as a result groundwater level is going down rapidly. Percentage of overexploited blocks in the country has increased drastically in the last two decades. Farmers are extracting groundwater without understanding the future impacts and as a result they are facing groundwater problems in many areas. Short term vision and interest in current benefits are the two major drivers for this unending quest of water. This has led to two basic problems; one, groundwater depletion and two, unequal distribution of water within the farming community. Old mechanisms of groundwater abstraction are no more in existence and big farmers, who have money to drill bore wells, are abstracting more water. Small farmers who are irrigating

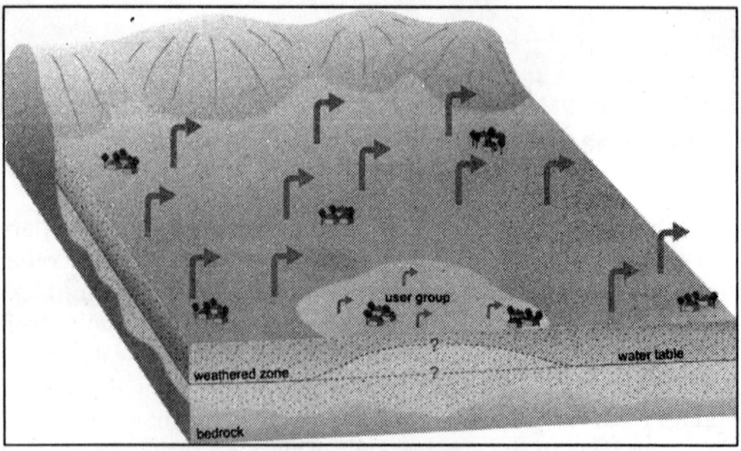

Figure 13.5: Free Riding on Groundwater Resource (COMMAN, 2005)

with dug wells are the main sufferers in this scenario. Water levels in their wells are going down because of pumping through bore wells in the nearby areas.

In areas where dug wells are the only source of drinking water, excessive pumping has also caused a potential threat to drinking water security in two ways: a) it has affected the water availability in the drinking water wells and b) new water quality problems like Fluoride, Arsenic and Iron contamination are arising because communities are shifting from wells to hand pumps or chapakals for drinking water. Hand pumps tap the deeper confined aquifers; this water is old and is in contact with the rocks for a longer period of time and there are grater chances of geogenic contamination.

At a local level, communities do not recognize groundwater as a common resource. Even in areas where communities have come together for surface water management, groundwater is utilized on individual basis. This mentality has encouraged free riding over groundwater resource. Communities undertaking recharge measures have to share their benefits with other communities, irrespective of their involvement in recharge programmes, since both communities are sharing the same aquifer. In this case, the second community is the free rider. In the absence of any management mechanism or long term utilization strategy, this inequity and threat to the groundwater resource will always remain high. There is an urgent need to intervene and evolve strategies of sustainable management of groundwater. Such strategies should involve local communities. Communities are the main stakeholder and actual users of groundwater and any intervention without involving the community will not yield any positive results. This will require building the understanding and capacities of the communities and finding the area specific solutions for local level problems. There is a need to start community trials in various diverse groundwater settings *i.e.* groundwater typologies to evolve a resource management strategy. Community level institutional mechanisms working with a supportive legal set-up are the keys for sustainable management of groundwater.

Groundwater Management Process

The idea of community based groundwater management in itself brings many questions to mind, the most basic being–*what is the starting point*? Groundwater management requires a multipronged approach and set of efforts at various levels like, at resource level, at community level and at governance level. These efforts have to go in tandem with each other and it also requires a clearly define strategy. Broadly we can divide this strategy into the following four areas:

Understanding the Resource

In most water managements drives, groundwater does not get the importance and space that it should get. In most watershed development and drinking water programmes, groundwater is considered as a sub activity and negligible effort is put in for its management. The main reason for this is the lack of information and understanding of this resource. Understanding the groundwater resource is perhaps the first step towards its management. Understating groundwater means understanding its occurrence, movement, availability, fragility and also the nature of aquifers. The science of hydrogeology gives us the tools to understand groundwater. Geological mapping and groundwater level data collection provide immense information for the mapping and delineation of aquifers and it should be done at a regional and local level for the better understanding of the groundwater systems and their recharge-discharge areas. Hydrogeology helps in estimating groundwater availability and safe limits of water abstraction from the aquifer. Understanding the resource also includes an understanding of groundwater quality. The nature of groundwater dependency and current as well as historic pattern of groundwater use at local and regional level are other factors which have to be studied. Sharing and percolation of knowledge at a community level is necessary for the community to understand the complexity

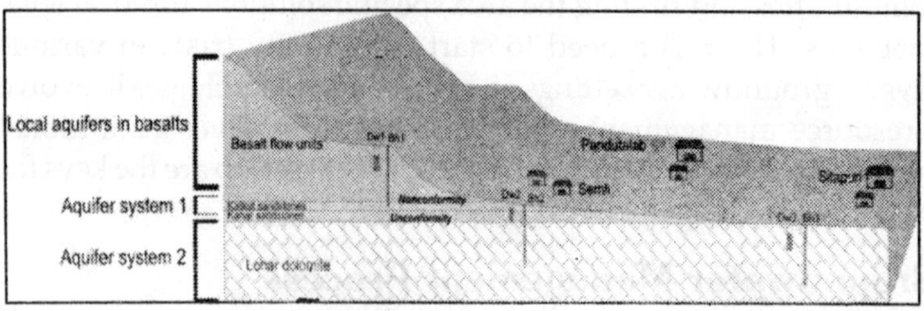

Figure 13.6: Aquifer Delineation at a Local Level

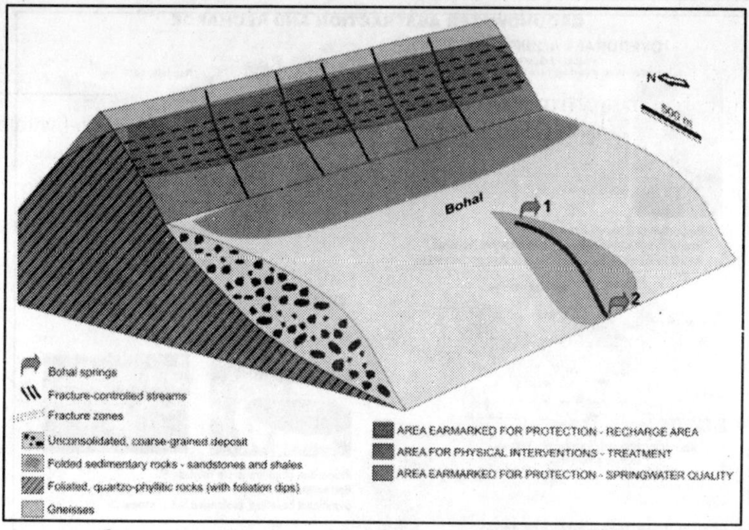

Figure 13.7: Geology and Management Strategy for Bohal springs, Palampur, H.P.

of the resource which is pre-requisite for its sustainable management.

Sustainable Utilisation and Resource Protection

Sustainable utilisation has two inbuilt aspects; a) effect of abstraction on overall health of the resource and b) long term availability of the resource for future generations. Unsustainable abstraction can put an adverse effect on aquifers and deplete them in the long run. A scientific understanding of aquifers tells us the safe limits of groundwater abstraction and measures to be taken for its protection. Aquifers have natural capacities to store (storability) and transmit (transmissivity) water. When pumping rate exceeds the rate with which water is coming to the well or bore well, it leads to unsustainable use of groundwater. Similarly, when abstraction exceeds the annual groundwater recharge amount it starts utilising the main storage of the aquifer and lead s to groundwater depletion. Density of wells/bore wells in an aquifer also determines the sustainable or unsustainable use of groundwater as pumping in one well will affect the water levels in nearby areas. Well density has both resource level and social implications, especially in the areas that are dependent on groundwater for their drinking water

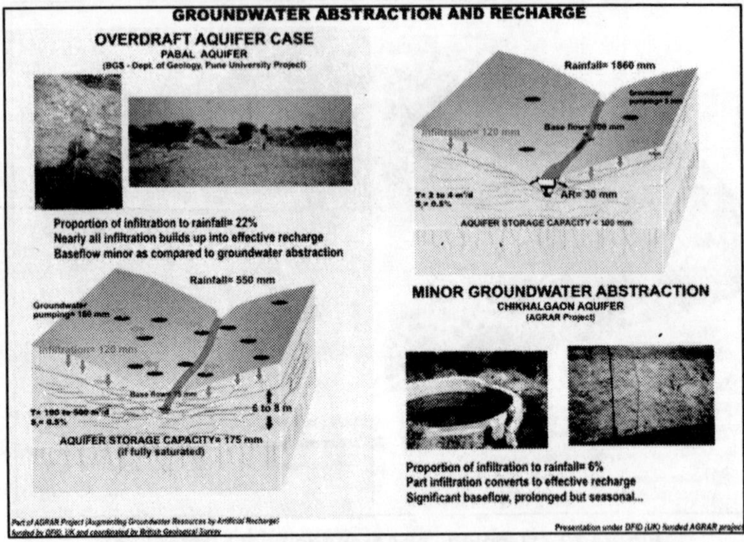

Figure 13.7: Understanding Recharge and Abstraction in Different Type of Aquifers (Kulkarni *et al.*, 2005)

requirements. Aquifer recharge is a natural process and it depends on the structure and nature of the aquifers, though artificial recharge measures can enhance the recharge to some extent. Protection of natural recharge areas is very crucial for the long term sustainability of the groundwater. Developing protocols, based on scientific study and developed in a participatory process, that emphasise resource protection, wise use and a community based management mechanism is the first step towards groundwater management. As the groundwater condition in India is very diverse, different sets of protocols are required for different groundwater typologies.

Ownership Related Issues

Current groundwater utilisation pattern attributes a very high degree of uncertainty with regards to the common ownership of the resource. *'Groundwater that is occurring under my farm is mine and all rights related to abstraction of that water also belong to me'* - is the general understanding of groundwater within the farming community. This sentiment has resulted in the overexploited state of groundwater that we see today. In this scenario establishing the acceptance of groundwater as a common pool resource is a big challenge and will require a series of efforts at various levels. First is, knowledge building about the resource which will require a sound scientific understanding of resource and also social skills to translate the scientific knowledge at local level. Second is building consensus in the community for common use of groundwater. This will require a community level formal declaration of *'only community use'* of groundwater. Third is developing practical mechanisms and protocols for common use of groundwater, which will define the rules, regulations and strategies of groundwater sharing.

ACWADAM has conducted community trials on the common use of groundwater and succeeded in three locations (Pondhe and Muthalne in Maharashtra and Nachanbore in Madhya Pradesh) where communities have declared groundwater as community resource and came up with following rules for groundwater use in their village:

- ☆ Groundwater will be considered as community resource and will be utilised by Water User Groups through community wells only.
- ☆ Private well digging and bore well drilling is banned in the village.
- ☆ All farmers in the village will irrigate equal areas of land and follow a common cropping pattern.
- ☆ Area of land to be irrigated and crop will be decided on the basis of rainfall and recharge of their aquifer.
- ☆ Well depth and distance between two irrigation wells as well as irrigation and drinking water well s be maintained according to hydrogeological guidelines.
- ☆ Common and regulated pump capacity is decided for pumping.
- ☆ Existing wells will either be converted into community wells or will be closed for irrigation use.
- ☆ Recharge area in the village will be protected and artificial recharge measures will be done by the community to enhance recharge.

Figure 13.8: Community Dialogue for Building Understanding and Consensus

Governance Related Issues

In the absence of any legal and institutional mechanisms, governance of any community based groundwater management effort at local level is a big challenge. In most of the states in India there are no Acts fcr groundwater management and in states where an Act has been created, it focuses on regulating authorities and regulation of groundwater abstraction in over abstracted areas. There is no scope for community management of groundwater in these Acts.

Historically seen pioneering efforts of groundwater management like Pani Panchayat are no more existent because of lack of appropriate legal backing. Today if one community is willing to come together for groundwater management and form Water User Groups, there is no legal support and backing for such efforts and scope of penalizing the offender or free riders. A very progressive and innovative act like "Maharashtra Management of Irrigation Systems by Farmers Act (2005)", which aims at community management of water resources, talks about surface water and is limited to canal irrigation projects only. In the absence of such an Act for Groundwater, Gram Panchayat can play a very crucial role and support village level institutions in groundwater management and governance at local level.

ACWADAM in its community based groundwater management trials involved Gram Panchayat as an active partner. Groundwater management plans and rules prepared by Water User Groups are passed in general meetings of Gramsabha and entered in the official proceedings, which gives them a legal validity. Gram Panchayat can use its legal powers to ensure the implementation of groundwater management rules and protocols made by Water User Groups and also penalise the offenders and free riders. In areas where there are standing committees for water resources these committee can also take this responsibility. This may not be a fool proof mechanism, as Panchayats today have very limited legal powers and other time-consuming responsibilities, but is the only option available at present. There is a need to bring groundwater under the purview of Acts like "Maharashtra Management of Irrigation Systems by Farmers Act (2005)" (for management purpose) or pass a similar Act that strengthens the community level groundwater management efforts.

Table 13.1: Introducing GW Management Protocols with Respect to Over Abstraction, in Different Phases of GW Use

Protocol	*Stage* 1*	*Stage 2*	*Stage 3*	*Stage 4*	*Stage 5*
Geo-hydrology in Watershed Development		✓	✓	✓	✓
Protection of recharge areas	✓	✓	✓	✓	✓
Efficient well use		✓	✓	✓	✓
Pump capacity regulation			✓	✓	✓
Distance (wrt* drinking water well) regulation			✓	✓	✓
Depth Regulation (wrt* drinking well)			✓	✓	✓
Regulation of Agricultural water use			✓	✓	✓
Groundwater management through sharing	✓	✓	✓	✓	✓

Groundwater development stages stated by Tushar Shah in his book- "Taming The Anarchy" ✓ Protocol to be introduced in the field.

*wrt: With respect to.

Conclusion

Today, groundwater is the main source of irrigation and in the future also it is going to be the main stay of Indian agriculture. Its management should be under a clearly defined social and scientific process and it should be contextualized according to the diversity in the resource and its utilization pattern. At the same time there have to be some broad guidelines for larger areas, with similar kinds of problems, which are based on field research, real time data and strong scientific backing.

ACWADAM has tried to understand India's groundwater related problems while working in different parts of the country and has also tried to propose situation specific solutions to these problems. These solutions are integrated in, a) protocols that ACWADAM has developed for groundwater over abstracted areas and b) process that is followed to develop the protocols and suggestions for implementation of protocols at field level. Groundwater management protocols deal with the broader level issues and need to be further simplified and elaborated according to the needs at local level. Protocol implementation will also be determined by the level or stage of groundwater development in that area. ACWADAM firmly believe that for reaching at a comprehensive groundwater management methodology there is a need for testing these protocols in many locations of different groundwater typologies.

References

COMMAN, 2005. Community Management of Groundwater Resources in Rural India, (Eds.) R. Calow and D. Macdonald. Research Report: British Geological Survey Commissioned Report CR/05/36N.

Freeze, R.A. and Cherry, J.A., 1979. *Groundwater*. Prentice-Hall Inc., Englewood Cliffs, New Jersey, p. 604.

Shah, T., 2008. Taming the Anarchy, Groundwater governance in South Asia Ostrom, E. Governing the commons, Cambridge, 1990.

Planning Commission, Government of India, 2007. Groundwater management and ownership: Report of the Expert group. (New Delhi: Planning Commission)

Chapter 14

Industrialization and Environment: A Sociological Analysis of ACC Barmana Cement Factory

☆ *Mohinder K. Slariya and Ashwani Sharma*

ABSTRACT

The present paper focuses on the other side of the development, based on the findings from the empirical study of ACC cement factory at Barmana situated in the lower Shivalik Himalayas in Himachal Pradesh. Like other developmental activities, this activity also ignores the ecological and environment concerns of the local area and disturbs the physical, socio-cultural as well as non-material environment. Along with this, this area is surrounded by Kol Dam, Dehar Hydel Electric Power Projects and Govind Sagar Lake, which are sufficient to create havoc anytime in this area. No doubt we need development, but the fundamental question remains unanswered, development at what cost. It is high time to check all such type of developmental activities by adopting human and environmental friendly technology. Development should be for man; man should not be for the development.

Keywords: Industrial development, Dams, Social impact, Culture.

Introduction

Hitherto all development has first taken place without concern for the environment. The reason being, that at that time the relationship between man and environment was limited only to the fulfillment of basic needs of the survival. It was initially the primitive man in neolithic stage of human existence, man first learnt to ignite fire and cultivate the fields and thereafter the desire of human beings to improve their existing conditions, put him on the road of development. With the introduction of modern technology the concept of development has become more crucial for the third world countries like India, who has under developed economy and technology. In order to achieve economic and technological development, India had to boost its industrial sector and it could have been done by only with the help of industrialization. Rapid industrialization has been one of the main instruments of achieving the goals such as removal of poverty and attainment of self-reliance. No doubt with these efforts economic conditions have improved, but at the same time it has caused considerable problems for the fragile environment by adding to pollution, soil erosion, displacement of people, health hazards and floods, which has been approved by the different studies done in this field. Industrialization implies setting up of more and more industrial units, which has immediate effects on the living pattern and traditional structure of society as well as of economy.

Himachal Pradesh being a mountainous state has a very limited agricultural area and difficult to modernize this sector. Therefore it is difficult to attain economic prosperity through this sector. Thus keeping in view this fact Himachal Pradesh has to plan and forgo ahead not only in agriculture sector but also in industrial sector. In this connection in 1962 Geological Wing of the industry department of Himachal Pradesh was setup with the objective to take intensive investigation and exploration for the limestone in Sirmour, Chamba and Bilaspur districts. The first mining lease for limestone was granted at Paonta Sahib in Sirmour district having an estimated reserve of 110 million tones and negotiated by the Cement Corporation of India. It was the first cement factory in Himachal Pradesh. The second factory was established in 1984 at Gagal (Barmana), which has been studied by the researcher, in Bilaspur district with a productive capacity of 600 tons per day. This area has cement grade limestone deposits of about 150 million tones, capable to meet the two million tones annual cement demand of the northern region of the country.

Table 14.1: Sample Selection of the Study

Barmana	15
Panjgawin	8
Toroe	12
Salaper	12
Dhamkoh	15
Total	60

Development is not an isolated concept; it is very much intertwined with environment. For any development, industrialization is needed and industrialization needs raw material from natural setting and place to build up infrastructure, which create disturbance in the physical surroundings of man. Developmental processes have generated enormous environmental damages in form of ecological degradation such as deforestation, desertification, pollution, loss of biological diversity etc. The rising environmental issues are challenging the theory and practices of development by bringing prosperity on the one hand and environmental problems on the others. In real sense instead of receiving human beings are paying more in terms of deteriorating quality of life and other environmental hazards.

This chapter is an attempt to understand the impact of industrialization on environment. It is based on the empirical findings from a cement factory located in lower hilly South Central Himachal Pradesh, with its headquarter at Bilaspur and situated at Shimla-Mandi National Highway-21. It is an attempt to gain insight into the environmental problems associated with industrial development at a micro level.

In order to analyze above said objectives five most affected villages (Barmana, Panjgawin, Taral, Salaper and Dhamkoti) situated in a semi–circle form have been undertaken for study purpose. These villages were largely affected by mining, blasting and allied activities. Out of 300 (approximately) affected households 60 households have been selected randomly by using simple random sampling method and exploratory research design has been used. The prime aim of the study is to explore the relationship between industrial development and environmental conditions of the area. Any descriptive and exploratory study depends upon two sources, namely Primary and Secondary. The former implies organization of fieldwork and latter refers to survey of literature. But due to the minimum availability of secondary sources on this topic, the primary has been mostly utilized.

Table 14.2: Change in Possession of Land

Land in Bighas	*Before Factory*		*After Factory*	
	No. of Responses	*Percentage*	*No. of Responses*	*Percentage*
Upto 10	8	13.33	7	11.69
11-20	18	30.00	22	36.67
21-30	19	31.67	17	28.33
31-40	7	11.69	5	8.33
41-50	4	6.63	1	1.67
51+	3	5.00	-	-
No Land	1	1.67	8	13.33
Total	60	100.00	60	100.00

Table 14.3: Impact of Occupational Profile of the Respondent

Study Sample	*Culti-vators*	*Govt. Employee*	*Factory Emplo-yee*	*Shop-keeper/ Business-man*	*Teachers*	*Total*
Cultivators	9*	1**	7***	1**	-	18 (30.00)
Govt. Employee	3**	-	-	-	-	3 (5.00)
Armed forces	3**	1**	1***	1**	-	6 (10.00)
Semi Govt.	2**	-	1***	-	-	3 (5.00)
Shopkeepers	-	-	-	1*	-	1 (1.67)
Students	1**	2**	19***	6**	1**	29 (48.33)
Total	18 (30.00)	4 (6.67)	28 (46.66)	9 (15.00)	1 (1.67)	60 (100.00)

Parenthesis indicate percentage.

*: No occupational change; **: Moved to other occupation; ***: Moved to factory.

This empirical study reflects into three following aspects; one; after the establishment of the factory a big and good part of village land has been acquired by the owner of the factory, two; new economic opportunities have been created in the term of jobs in factory, boost to local economy and factory's participation in developmental activities, three; a new category of people, the immigrants has been added to village population which positively and negatively affects the locality. The economic and social consequences of all these aspects under various heads have been examined, by using Structured Interview, Participatory Observation and Simple Random Sampling Method.

Table 14.4: Socio-Economic Background of the Respondents

	Study Sample	*No. of Responses*	*Percentage*
Social segment	Brahamin	10	16.67
	Rajput	25	41.67
	Khatri	3	5.00
	SCs	22	36.66
	Total	60	100.00
Age of respondent	Upto 30	23	41.66
	31-40	13	21.67
	41-50	10	16.67
	51-60	6	10.00
	61 and above	6	10.00
	Total	60	100.00
Educational profile	Illiterate	9	15.00
	Primary	4	6.67
	Middle	11	18.32
	Matriculation	19	31.67
	Intermediate	10	16.67
	Graduate	5	
	Post-graduate above	2	
	Total	60	100.00

There is very big difference between household amenities, occupational and income status and land holding status before and after the establishment of factory. Before the establishment of factory the conditions of the physical environment was very good, having a thick cover of forest of local trees, bushes and very good fertile land at factory site and its adjoining areas. But with the establishment of factory all the trees and bushes were cut down, which has given rise to the problem of soil erosion, clean air, fuel wood, fodder etc.

In order to meet the infra-structural need of the factory, the land of many people was acquired, which rendered many people displaced and compelled to migrate to other villages, other districts and states. With the coming up of the factory the problem of air, water and sound pollution has become acute, which has given rise to a number of diseases *i.e.* sore eyes, breathing, kidney and gall-bladder stone, allergic reactions, TB, impaired hearing, etc. The relationship between industrial development and environment is sociologically very significant because it not only affects the individual and household but the entire way of living. People were compelled to be displaced because there was dust from chimneys and blasting which negatively affected the productivity of land, which further led people to change land-using pattern. The burning of coal has adversely affected climatic conditions of the area. The problem of housing for the outside workers led to the problem of slum formation.

The Non-Material Environment of the area has been adversely affected after the establishment of factory. The impact of the industrial development on the traditional social structure, inter-caste relationship and cultural pattern of the area can largely be seen. Prior to factory people's outlook was very narrow, cultural pattern was traditional and inter–caste relations were rigid but with establishment of factory the whole scenario has changed.

Cultural patterns including dress, habits and ways of life have undergone change, inter-caste relationship has been changed. The whole area gives the impression of planned city having city like characteristics. Traditional joint family system has changed, now people prefer to nuclear families. Beside these, the factory has contributed a lot in the boosting of local economy by proving direct and indirect employment to a number of people. There is English Medium School, arrangement of sanitation, footpath, shopping complex, Senior Secondary School and ACC Government Degree College at Ghumarwin.

In the light of foregoing description it can be concluded that on the one hand it is bringing prosperity, but on the other it is affecting the life of natives adversely, they are compelled to pay a huge cost of these developmental activities. Instead of other side of development it is recognized fact that the industrialization is the need of the hour. It is backbone of the development, in absence of which no any developmental activities can take place. No doubt we need development, but at what cost? It should not at the cost of human survival. We have to adopt such techniques through which the ill effects of developmental activities can be minimized. We have to adopt the strategy through which displacement of people, social and cultural environment of the area can be saved. We should not forget that we all are the constituents of the delicate web and each organ has an important role to play to maintain the web. A little trespass even in the name of development may put a question mark on our survival. *Development should be for man; man should not be for the development*. All life on earth is inter-dependent and man is only a little strand in this delicate relationship.If man and nature do not find ways and means to live together, ultimately there will be no life on the earth. *With a new hope for the future in which all the developmental activities may be eco-friendly* In the end it can be stated that

Nature can survive without man, but man cannot.

Chapter 15

Climate Change and its Impact on Seasonal Streams Flowing in the Natural Forest Areas of Central India

☆ *Manish Mishra*

ABSTRACT

Availability of freshwater is an increasing concern across the country and it may be the most important natural resource issue of the decade. Climate change and its effects on water are expected to intensify freshwater scarcity and conflict. Climate change affects not only the quantity of surface water but also quality of surface and groundwater. Warming over the past decade has fundamentally altered the hydrologic cycle and these changes are percolating through watersheds. Precipitation pattern is changing and repeated floods and draughts, adversely impacting the seasonal stream flow and water availability. The storage and availability of drinking water and flow in streams in the *natural forests* is adversely affecting the flora and faunal habitat, wildlife movements etc.

Madhya Pradesh is one of the richest states in terms of natural forests and bio-resources. Most of the tribal peoples of central India are residing around the forests and depending upon the seasonal streams flowing in their vicinity. Due to continuous draughts and less rainfall, they are facing acute shortage of water. Most of the seasonal streams were dried during summer (*i.e.* February to June) which earlier flows round the year. The tribal women are forced to go minimum 3-5 km. in search of water for their daily needs. The cases of tribal people and wildlife conflicts are rising as the wild animal in search of water entering to village settlements, more frequently. The paper discusses adverse impact of climate change on seasonal streams and coping strategy of Tribal of Central India. The paper highlights the importance of conserving natural flowing streams located in the forests to provide clean, abundant water.

Keywords: Tribal, Water, Streams, Draught, Flora.

Introduction and Review of Literature

Worldwide changes in seasonal patterns, weather events, temperature ranges, and other related phenomena have been reported and attributed to global climate change. Numerous experts in a wide range of scientific disciplines have warned that the negative impacts of climate change will become much more intense and frequent in the future particularly if environmentally destructive human activities continue unabated. India, whose climate is largely controlled by an annual monsoon, appears to be experiencing increasingly severe and erratic precipitation. A recent study found that the overall amount of monsoon rainfall across central India has remained relatively stable over the past century; however, moderate rainfall events during monsoon seasons have significantly decreased while extreme rainfall events have greatly increased since the early 1980s. This increase in extreme rainfall events could indicate greater potential for future natural disasters (Cavaliere, 2009).

Forests play key roles in supporting national economic activities and providing livelihood portfolios for many in India. They provide valuable ecosystem services like climate regulation, hazard protection, water conservation, and also affordable goods like fuel wood, foods and nutritional supplements, and medicinal products etc. most of which depend on biodiversity. They are at the frontline in mitigating climate impacts on India by reducing exposures to scourging heat, dust storms and floods. In spite of the frequency of drought disaster in several parts of India, there are no national or regional institutional structures for climate change adaptation, and for biodiversity conservation except in designated protected areas, like there is for food and water in response to drought and desertification. The role of seasonal; streams flowing in the natural forest areas and their conservation is largely ignored by the planners. Climate change and its impact on availability of water in watersheds and its impact on locals remain insufficiently addressed.

There is large body of knowledge and experience within local communities on coping with climatic variability and extreme weather events. Local communities have always aimed to adopt variations in their climate. To do so, they have made preparations based on their resources and knowledge accumulated through experience of past weather patterns. This includes times when they have also been forced to react to and recover from extreme events such as floods, drought and hurricanes. Local coping strategies are an important element of planning for adaptation. Climate change is leading communities to experience climate extremes more frequently, as well as new climate conditions and extremes. Traditional knowledge can help to provide efficient appropriate and time tested ways of advising and enabling adaptations to climate change due to global warming. Several example of local coping strategy are mentioned in the background papers of the workshop (UNFCC 2006,2007a, 2007b).

In Latin America, local coping strategies include a variety of agricultural practices, ecosystem protection, and methods to adapt to extreme events. Farmers in Peru have been using an ancient irrigation and drainage system Waru-Waru, or raised field agriculture, which makes it possible to bring into production into low lying, flood prone, poorly drained lands. The shallow canals provide moisture during draughts and drainage during the rainy season. When filled with water they also create a microclimate that acts as a buffer against night time frosts. The Aymaran indigenous people of Bolivia have been coping with droughts through the construction of small dams"qhuthanas". Coping strategies include agricultural techniques, plant/animal protection and coping with climate extremes. (UNFCC, 2007b).

In Asia rural farmers have been practicing a range of water storage techniques as coping strategies and tactics to enable sustainable food production and deal with extreme events. Various water use and conservation strategies include terracing, surface water and ground water irrigation and diversification in agriculture to deal with drought. Structural and non-structural measures are used to deal with flood and coastal inundation. Changes in temperature and precipitation will continue to decrease snow pack, and will affect stream flow and water quality throughout the Pacific Northwest region.

Water quantity and quality have become serious issues facing many communities and nations around the world following the changes in climate and fast rising human population (Kundzewicz *et al.*, 2007). Addressing these issues requires knowledge of how water resources are affected by changes of various aspects of regional hydrological cycle. Inquiries of such knowledge have been the theme of many studies that examined climate and human induced effects on hydrological cycle of various spatial and temporal scales. These studies revealed some interrelationship of climate and land-use changes with various aspects of regional hydrological cycle (*e.g.*, Dunn and Mackay, 1995; Mander *et al.*, 1998; Krause, 2002; Ren *et al.*, 2002; Tao *et al.*, 2003; Twine *et al.*, 2004; Hu *et al.*, 2004). Results of these studies show varying effects of land-use and land-cover on surface stream flow and complications of such effects in different climatic conditions. For example, Lahmer *et al.* (2001) show that in wet climate regions even some extreme land-use change only resulted in comparatively small impacts on regional water balance, a result consistent to the finding by Legesse *et al.* (2003) for tropical Africa. Dunn and Mackay (1995) further show that variation in the water balance resulting from land-use change may have different effects on the hydrology, depending on the nature of the soils, revealing an important role of the soils in regional water cycle change. Interferences to these changes by policy and management decisions also have been examined to provide insights for effectively mitigating negative effects of climate change on water resources (*e.g.*, Ren *et al.*, 2002; Mimikou *et al.*, 1999).

This vulnerability has been further elevated by deforestation and change of landscape in the basin. Although forests in the basin recovered to nearly 60 per cent during the 1990s after changes of government policy and enforcement of local conservation measures, the emphasis on economic returns (values) and subsequently plantings of spruce and pine instead of local/native plants have changed the population and composition of the basin's land-cover.

These changes in land-cover and vegetation have affected the surface and groundwater hydrology and stream flow in the sub-basins of the Poyang Lake, altering the hydrological cycle and flood vulnerability of the lake basin (Hu, 2001). In addition, these effects vary as functions of seasonality and the changing climate (Huxman *et al.*, 2005). To understand the causes of these variations, it is necessary to understand stream flow responses to changes in land-use/land-cover and climate. Knowing these responses we can address the questions of how the on-going land-use change may have influenced the annual and seasonal stream flow, and flood potential under the current climate, and how such influences may change with future climate changes. Answers to these and related questions will improve the predictability of hydrological consequences of land-use and climate change. This ability is essential for long-term planning of land-use to not only protect water resources but also effectively manage floods as well as droughts.

Material and Methods

Study Site

Mandla, a tribal district of Madhya Pradesh, is situated in the east-central part of the state. Mandla is located at 22°362 N 80°232 E/22.6°N 80.38°E/22.6; 80.38. It has an average elevation of 445 meters. The district lies almost entirely in the catchments of river Narmada and its tributaries. A district with a glorious history, Mandla is endowed with rich forests. The town is picturesquely situated in a loop of the Narmada River, which surrounds it on three sides, and for 15 miles between Mandla and Ramnagar the river flows in a deep bed unbroken by rocks. Once this small town was considered among the greenest towns; now this is just a memory.

Table 15.1: List of Surveyed Forest Ranges, Villages of Mandla Forest Division, Madhya Pradesh

Name of Range	*Name of Selected Village*	*Household Surveyed*
Motinala	Baila, Tatma, Nonadar, Devgaonv, Chemaghundi	20 per cent
Mawai	Sathiya, Sajalagan, Patpara, Amwar, Awlaghughra	20 per cent
Anjaniya	Ghot, Medhatal, Magdha, Umariya, Sajpani	20 per cent

Figure 15.1a: Map Showing Madhya Pradesh and Mandla Districts

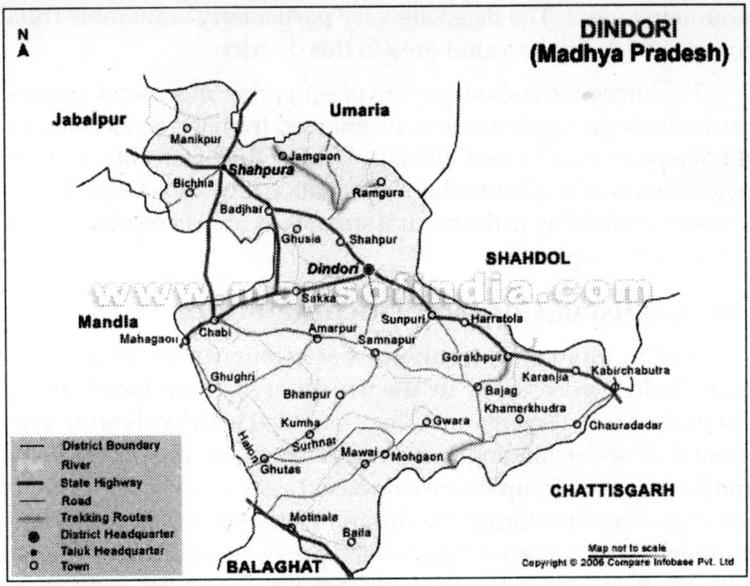

Figure 15.1b: Map Showing Dindori Districts of Madhya Pradesh

Deforestation has largely affected this area as shown in summer 2007 when the temperature exceeded 46°C.

The district of Dindori encompasses a total area of 571883 hectares. It is located at latitude 22°.17 to 23°.22N longitude 80°.35 to 80°.58 E. The district comprises of 97 per cent rural and tribal population. The district draws its revenue from agricultural and forest products. Dindori has not yet experienced industrialization. The holy river Narmada passes through the district. It is situated at a height of 1100 m above sea level amongst herbal-rich, Maikal

Table 15.2: List of Surveyed Forest Ranges, Villages of Dindori Forest Division, Madhya Pradesh

Name of Range	*Name of Selected Village*	*Household Surveyed*
Bajag	Khamera, Chadha, Tatar, Silpidi, Tarach	20 per cent
Samnapur	Dhaba, Paudi, Ajgar, Kandavani, Dhurkutta	20 per cent
Karanjia	Kandatola, Ladradadar, Kharidih, Padripani, Kindrabehra	20 per cent

mountain ranges. The Baiga are very particularly vulnerable tribal groups which can be found only in this district.

The information about pattern of water flow in seasonal streams particularly during summer was obtained from the local villagers, old age person and forest officials. During field visits of the study area efforts were made to collect information regarding timeline data of water availability in the natural streams located in nearby forests, by the local people.

Group Discussion and Documentation

The random questionnaire based survey of 20 per cent households were done in the study area. Few focus group discussions was also organized for discussing with local community about their understanding on seasonal stream flow, presence of water and pattern etc. Group discussion based assessment with regard to water availability during the different seasons and flow of water during the year, for before 50 years and present timeline was collected with the help of middle aged, old aged and younger generation. Documentation on experiences of local community for the present condition of seasonal streams and reasons for its dryness was also gathered.

Results

The data presented in Table 156.3 shows seasonal pattern of natural stream water flow (nallah) during last fifty years in the natural forests of Mandla and Dindori forest divisions. During group discussion it was opined by the villagers that in both districts earlier (fifty year back or so), plenty of water is flowing in all the natural streams during the year. In the past twenty years, water flow in streams for 6-8 months in a year in Mandla and 8-9 months in Dindori division. However, currently water flows only 3-4 months in a year in the natural streams of Mandla and 4-5 months in Dindori forest division.

Data depicted in Table 15.4 shows availability of water in the natural streams of both the selected forest division/ranges. On an average fourteen seasonal streams flows in Mandla out of which 9.0 (64.28 per cent) were dried during summers. Only two streams (14.28 per cent) were found having water during summers in Mandla, whereas water is available in six streams (30 per cent) in Dindori

Table 15.3: Pattern of Seasonal Stream (Nallah) Water Flow during Whole Year in Mandla and Dindori Forest Divisions

Sl.No.	*Name of Forest Division*	*Pattern and Availability of Flowing Water in Seasonal Streams*		
		*Stream Flow Before 50 Years**	*Stream Flow Last 20 Years**	*Stream Flow Now*
1.	Mandla	Whole year	6-8 months	3 to 4 months
2.	Dindori	Whole year	months	4 to 5 months

* Information provided by villagers of study area.

forest division. Maximum 16 (80 per cent) streams were found dried during summer in Motinala range, whereas minimum seven (50 per cent) were found in Mawai range of Mandla division. On the other hand twenty seasonal streams were flowing in the natural forests of Dindori division, out of which more than ten were dried (52.5 per cent) during summers. Maximum 08 (66.5 per cent) streams were found dried during summer in Karanjia range, whereas comparatively minimum nine (50 per cent) were found in Samnapur range of Dindori division.

Discussion

Availability of water and quality has become serious issues facing many communities and nations around the world following the changes in climate and fast rising human population. Deforestation and habitat fragmentation has triggered the problem particularly in remote forest villages of Central India. A large number of local communities were depending upon seasonal streams flowing in the dense forest areas and local vicinity. These natural streams not only provide drinking water but simultaneously used for *nistar* (bathing, washing etc.) purposes. The non-timber forest products (NTFPs) collectors' utilized stream water for cleaning, washing and boiling of wild fruits, roots, medicinal plants, barks etc. Besides this these are the source of drinking water for wild animals, cattle etc. More presence of seasonal streams in the forests indicates that soil moisture level and ground water table is high in that particular locality. However, in the natural forests, the presence of seasonal streams was more in hilly areas than plains.

Table 15.4: Seasonal Pattern of Flow of Water in Seasonal Streams (Nallah) Located at Mandla and Dindori Forest Divisions

Name of Forest Division	*Name of Forest Range*	*Total No. of Flowing Streams in Each Forest Range*	*Stream Flow having Water (Wet) during Summers*	*Stream Flow having Stable Water (in Patches) during Summers*	*Stream Flow having no Water (Dried) during Summers*
Mandla	Motinala	20	01(05.0%)	03(15.0%)	16(80.0%)
	Mawai	14	02(14.5%)	05(35.5%)	07(50.0%)
	Anjaniya	08	01(12.5%)	01(12.5%)	06(75.0%)
Average of division		14	2.0(14.28%)	3.0(21.42%)	9.0(64.28%)
Dindori	Bajag	27	07(26.0%)	05(18.5%)	15(55.5%)
	Samnapur	21	07(33.3%)	05(24.0%)	09(42.5%)
	Karanjia	12	02(16.5%)	02(16.5%)	08(66.5%)
Average of division		**20**	**06(30.0%)**	**04(21.5%)**	**10.5(52.5%)**

During group discussion it was observed that in study area, earlier (fifty years back or so), plenty of water is flowing in all the natural streams round the year. However, currently water flows only 3-4 months in a year in the natural streams of Mandla and Dindori forest division. This shows sharp decline of water availability in seasonal streams particularly in summer months. On the contrary, out of 14 seasonal streams flows in forests of Mandla a majority 9.0 (64.28 per cent) of streams were dried during summers and only two were found having water during summers. On the other hand, water is available in six streams (30 per cent) in Dindori forest division round the year whereas more than 50 per cent were found dried during summers. As a result, seasonal stream flow timing will likely shift significantly in sensitive watersheds (Littell *et al.*, 2009). Similar studies were also done by Mander *et al.*, 1998; Hua *et al.*, 2008, which shows varying effects of land-use and land-cover on surface stream flow and complications of such effects in different seasons.

Now due to climate change, reduction in forest cover and degradation of habitat, the occurrence of natural seasonal streams is sharply declining. Presence of water and flow of water in seasonal streams round the year is now a rare phenomenon in the study area. Most of the seasonal streams emerges during rainy season and disappeared after few months. However, due to less rainfall and subsequent draughts, the condition is more precarious in the study area. Presence of water in these streams during whole year is now a rare phenomenon chiefly due to population increase and deforestation. Villagers particularly women are the most sufferers due to shortage of water in seasonal streams. They have to travel at least 5-6 kilometres in search of water, which is earlier available plenty in their close vicinity or in nearby forest areas.

Conclusions

Results of this study show that climate plays a leading role in changing stream flow in the natural forests of Mandla and Dindori forest divisions of Madhya Pradesh state. A specific finding of this study is that the seasonal variations of water availability in different seasons over last fifty years and water flowing pattern in seasonal streams is showing sharp decline due to climate change and deforestation. More than 70-80 per cent streams were dried particularly during summers while earlier (before 50 years) they were flowing round the year. Different land-cover patterns/settings in the natural forests of selected districts can cause increases or decreases of stream flow in different seasons. This finding indicates that through interactions of vegetation with soil and surface swaters land-use and land-cover can modify the local hydrological cycle and exert a strong biological control on stream flow in the natural water streams. This finding further points to a potential that water availability for specific season(s) may be affected by continuous draughts and reduction in forest cover, degradation of habitat etc. This potential could be particularly attractive for tropical climate regions where precipitation occurs in rainy and rainfall decreases substantially in the summer growing season. Some land-cover change occurring simultaneously with an increasingly dry climate would further reduce stream flow. These results suggest that certain land-use and land-cover changes can modify drought intensity as well as frequency in the dry climate. An important implication of these findings is that it could be possible to mitigate undesired climate change effects (*e.g.*, intense floods or droughts) on environment and water resources by planning land-use to achieve specific hydrological effects of land-cover in the natural forests of study area. Effective and efficient control over deforestation and protection of natural habitats are the measures to improve soil moisture condition, stopping erosion and recharging of water in the natural seasonal streams (nallah) found in the forests.

Acknowledgement

Authors are thankful to Dr. R. B Lal, Director, for his kind inspiration and Regional Centre of National Eco-development Board, IIFM, Bhopal for providing financial support.

References

Courtney Cavaliere, 2009. The effects of climate change on medicinal and aromatic plants. Herbal Gram. *The Journal of the American Botanical Council*, 81: 44–57.

Dunn, S.M. and Mackay, R., 1995. Spatial variation in evapotranspiration and the influence of land use on catchment hydrology. *Journal of Hydrology*, 171: 49–73.

Hu, X.Y., 2001. Ecological and environmental changes in the Poyang Lake basin in the last century. *Bull. Jiangxi Normal Univ.*, 22: 365–370 (in Chinese).

Hu, Q., Feng, S., Guo, H. and Jiang, T., 2007. Interactions of the Yangtze river flow and hydrologic processes of the Poyang Lake, China. *Journal of Hydrology,* 347: 90–100.

Hu, Q., Willson, G.D., Chen, X. and Akyuz, A., 2004. Effects of climate and landcover change on stream discharge in the Ozark highlands, USA. *Environ. Model. Assess.,* 10: 9–19.

Hua Guoa and Qi Hu Tong Jiang, 2008. Annual and seasonal streamflow responses to climate and land-cover changes in the Poyang Lake basin, China. *Journal of Hydrology,* 355: 106–112.

Huxman, T.E. *et al.*, 2005. Ecohydrological implications of woody plant encroachment. *Ecology,* 86: 308–319.

Krause, P., 2002. Quantifying the impact of land use changes on the water balance of large catchments using the J2000 model. *Phys. Chem. Earth,* 27: 663–673.

Kundzewicz, Z.W., Mata, L.J., Arnell, N.W., Doll, P., Kabat, P., Jime´nez, B., Miller, K.A., Oki, T., Sen, Z. and Shiklomanov, I.A., 2007. Freshwater resources and their management. In: *Climate Change 2007: Impacts, Adaptation and Vulnerability,* (Eds.) M.L. Parry, O.F. Canziani, J.P. Palutikof, P.J. van der Linden and C.E. Hanson. Contribution of Working Group II to the Fourth Assessment Report of the Intergovernmental Panel on Climate Change. Cambridge University Press, Cambridge, UK, pp. 73–210.

Lahmer, W., Pfiitzner, B. and Becker, A., 2001. Assessment of land use and climate change impacts on the mesoscale. *Phys. Chem. Earth (B),* 26: 565–575.

Legesse, D., Vallet-Coulomb, C. and Gasse, F., 2003. Hydrological response of a catchment to climate and land use changes in Tropical Africa: Case study South Central Ethiopia. *Journal of Hydrology,* 275: 67–85.

Littell, J.S., Elsner, M. McGuire, Binder, L.C. Whitely and Snover, A.K. (Eds), 2009. The Washington Climate Change Impacts Assessment: Evaluating Washington's Future in a Changing Climate, (PDF 14.1 MB) Climate Impacts Group, University of Washington, Seattle, Washington.

Mander, U., Kull, A., Tamm, V., Kuusemets, V. and Karjus, R., 1998. Impact of climatic fluctuations and land use change on runoff and nutrient losses in rural landscapes. *Landscape Urban Plan.*, 41: 229–238.

Mimikou, M.A., Kanellopoulou, S.P. and Baltas, E.A., 1999. Human implication of changes in the hydrological regime due to climate change in Northern Greece. *Global Environ. Change*, 9: 139–156.

Ren, L., Wang, M., Li, C. and Zhang, W., 2002. Impacts of human activity on river runoff in the northern area of China. *Journal of Hydrology*, 261: 204–217.

Tao, F., Yokozawa, M., Hayashi, Y. and Lin, E., 2003. Future climate change, the agricultural water cycle, and agricultural production in China. *Agri. Ecosys. Environ.*, 95: 203–215.

Twine, T.E., Kucharik, C.J. and Foley, J.A., 2004. Effects of land cover change on the energy and water balance of the Mississippi River basin. *J. Hydrometeor.*, 5: 640–655.

UNFCC, 2006. Report on African Regional Workshop on Adaptation, Accra, Ghana, 21–23 September,2006.

UNFCC, 2007a. http://unfccc.int/files/essential_background/background_publications htmlpdf/application/txt/pub_07_impacts.pdf

UNFCC, 2007b. Report of the Conference of the Parties on its thirteenth session, held in Bali from 3 to 15 December 2007. Part One: Proceedings.

Chapter 16

Water Management Strategies for 21st Century

☆ *Sapna Bakshi and Vikas Bakshi*

"Water, Water everywhere,
Nor any drop to drink."

You all might have heard the famous lines of the poem, "The, Rime of The Ancient Marnier", written by S.T. Coleridge in 1798. In these lines, the poet has described the plight of the sailors on a ship lost in the ocean. I want to describe in my paper that water is very precious and a drop of water is worth more than a sack of gold to a thirsty man. Planet earth has abundance of water. Water has, indeed, played a significant part in shaping the history of mankind. This colourless and tasteless liquid is in fact so valuable that it deserves to be called the elixir of life. Circulation of fresh water through natural hydrological cycle available to human being for any use is a very small fraction (0.3 per cent) of the total world's water (1.36×100000000 M ham). This fresh water stock, which was almost constant, is now increasing due to glacier melting and climatic change because of global warming. This century is going to face many challenges due to global warming. In the past 100 years world's population has increased three times resulting increase in water use 6 times. It seems that the greatest scarcity in the 21'st century will be of fresh water, not of oil and fossil fuel.

There is no dearth of water in India. Our country receives adequate fresh water annually to grow more than two good crops on its entire cultivated area through integrated water management bringing 200 M ha gross cropped area under irrigation ensuring food security in the 21'st century and maintaining balanced ecosystem. To achieve this we have to shift from sectoral to integrated approach in water resource development and management. Our nation has been facing severe floods and drought problems very frequently and there is no reliable long term forecasting for these events. After occurrence of flood/drought we start thinking and discussing about its control measures and contingency plans but land up with only relief measures. Now we have to develop preventive measures for flood and drought mitigation measures to control these events permanently.

There has been a steady development of irrigation in India. The sectoral irrigation development projects in the past have increased our food production significantly but we can not go now in such sectoral way of irrigation development projects because of their adverse effects like water logging, salinity, lowering ground water table, displacement of people, silting of reservoirs and uneven water distribution. Moreover all good sites of major irrigation projects are built up and available sites are going to face challenging and difficult problems. National Water Development Agency's proposal of linking Indian rivers is currently under hot discussion and the task force on inter-linking of rivers, studying feasibility of various links, is now facing a lot of problems. Ongoing proposal of linking rivers is again designed as sectoral project and doubts are raised on its high cost, low efficiency, poor sustainability, imbalance ecosystem, inter-state dispute etc. Ultimately we have to link our rivers to transfer water from excess to deficit areas but before that we have to manage rainwater on watershed basins covering the whole basin areas. It may also be advised to consider following two points in this regard:

1. Global warming may increase flow of our snowfed rivers especially Ganga and Brahmaputra upto some extent during current century due to glacier melting and some of our cultivated areas near coasts and islands may come under sea water.

2. It is also speculated that China may divert upper catchment water before reaching to Indian Himalayan Rivers to its country reducing flow of Ganga and Brahmaputra by more than 20 M ha m.

There is only one way to solve these problems of drought, flood, low irrigation water availability and low productivity of land and water and that is systematic and scientific management of water on watershed basins covering whole basin area of rivers allowing adequate water to be conserved/stored where it falls. It requires following steps:

1. Store and conserve maximum rainwater where it falls on/in soil profiles, aquifers and in surface pond up to their available capacity.
2. The excess rainfall (runoff) if any should be stored and managed in gully/stream or in big tanks/reservoirs on watershed basins covering the whole basin following watershed principles for irrigation and groundwater recharge.
3. After adoption of steps 1and 2 controlled flood flow with less sediment during monsoon and more sub-surface base flow are expected in streams and tributaries of rivers. Small, medium and major surface irrigation projects can easily be taken up with less cost, more capacity and long life.
4. After adoption of steps 1, 2 and 3 the river flow will get regulated and changed. There will be less monsoon flood flow and sediment reducing chances of flood down stream and more base flow during post monsoon period providing more clean water in rivers.
5. After adoption of steps 1, 2, 3 and 4 there may be regulated and controlled flow in the down stream of the river and a part of which can easily be diverted to deficit area by linking rivers. Some river flow must be left to join sea for maintaining eco-system.

Above steps will provide a series of small, medium and large tanks/reservoirs and flowing water in streams, rivulets and rivers. Before using the stores (tanks/reservoirs) and flowing water for irrigation must be used for aquaculture. These steps will provide more recharged ground water which can be used for irrigation through shallow and deep wells for irrigation. Each well/tube should have fish pond and irrigation water should flow through fish pond. In addition to multiple use of irrigation water for fish and crop production, conjunctive use of rainwater, surface water and ground water should be followed in all cropped lands in the whole basin on watershed basin whether it is dominated by canal, well or pond commands. These steps indicate that integrated watershed development and management should be a pre-requisite of linking rivers and where these steps are taken on river basin level; watersheds will include all rainfed areas, water resources and their command areas. The state agricultural universities and Indian Council of Agricultural research have generated a lot of information and viable packages of practices for land and water management can be developed and used to saturate the whole basin. This water management system will solve most of the problems in following ways:

1. A bottom-up approach starting from individual field ending at river linking.
2. Farmers, villages, panchayats, districts and states will be getting full chance to store and conserve adequate water according to their needs and capability. Water sharing disputes will be minimized
3. Floods and water logging can be reduced and controlled permanently.
4. Irrigated command area and cropping intensity can be increased.
5. Irrigation systems will be more effective and sustainable.
6. Water productivity can be increased through multiple and conjunctive use of water
7. Provide balanced eco-system.

India has been primarily an agricultural country and still farming continues to be the main occupation of its people. At the end of 20th century the net and gross cropped areas were 142 and 190 M ha and irrigated areas were 57 and 79 M ha respectively. There is no dearth of water for irrigation to meet food requirement of our growing population in the 21st century but it requires integrated development and management of all the available water on watershed basins saturating the river basins and transfer of excess water to deficit areas through linking rivers.

There is a lot of wastage of water at domestic level and in irrigation. The efficiency of our irrigation system is very low, canal 38 per cent and ground water wells 60 per cent. In many areas water is available to users at no cost or at very heavily subsidized price. Thus, neither water managers nor users have incentives to conserve water, so water is over-used and wasted instead of being treated as a scarce resource. The answer is adequate pricing of water reflecting its true cost. Once, it is adequately priced people will conserve it and will know its importance.

Thus, to conclude, though water is the commonest of liquids, its amazing properties make it a unique resource. Investigations into the nature and properties of water could constitute an area of great scientific interest and is far from an exhausted field of research.

Chapter 17

World's Largest Canal based Drinking Water Project

☆ *S.M. Yadav*

ABSTRACT

A concrete gravity dam, 1210 meters in length and with a maximum height of 163 meters above the deepest foundation level, is under construction across river Narmada, Gujarat, India. The dam will be the third highest concrete dam (163 meters) in India. This dam will be ranking as the second largest in the world with an aggregate concrete volume of 6.82 million cu.m. The first is Grand Coule Dam in USA with a total concrete volume of 8.0 million cu.m. This dam with its spillway discharging capacity of 87,000 cumecs (30.70 lac), will be the third in the world, Gazenba (1.13 lac cumecs) in China and Tucurri (1.0 lac cumecs) in Brazil being the first two. The gross storage capacity of the reservoir is 0.95 M. ha.m. (7.7 MAF) while live storage capacity is 0.58 M.ha.m. (4.75 MAF). The dead storage capacity below minimum draw down level is 0.37 M. ha. m. (2.97 MAF). The construction of dam up to 121.6 meter is over and construction of 456 km long main canal is almost completed. Out of the total water resources of the state (Ground as well as surface water) the areas of Kachchh, North Gujarat and Saurashtara have 2 per cent, 11 per cent and 16 per cent water resources respectively which covers 58 per cent area of the state. These areas are subjected to draught having once in a five year frequency. These areas are one of the most water stressed area of the state as well as country. The mass migration tendencies seasonal as well as permanent towards Central and South Gujarat are observed regularly due to shortage of water, food and fodder. The main canal of Narmada dam has become new source of water for this region. The Sardar Sarovar Canal based drinking water supply project is the canal based biggest drinking water supply project in the world. It will cover Saurashtra, Kachchh and parts of North Gujarat. The progress so far is 68 per cent of villages are covered against proposed coverage. The project includes 48 bulk schemes, 123 distribution groups and a 3,250 km long pipeline network. The project is extremely important for Gujarat, as it will cover 53.3 per cent villages and 54 per cent urban centers of the state. It will cover 100 per cent villages in Ahmedabad, Amreli, Bhavnagar, Gandhinagar, Jamnagar, Junagadh, Kachchh, Porbandar and Surendranagar districts. In this project, water will be transported through bulk water transmission pipelines and distribution network to the doorsteps of villages and towns. The world's largest canal based drinking water supply project with its problems and remedial measures are discussed in this paper.

Keywords: *Water resources, Narmada main canal, Drinking water, Issues, Sustainable development.*

Introduction

The state of Gujarat can be divided into four parts as far as available water resources are concern (*i*) South and Central Gujarat (*ii*) Saurashtara (*iii*) North Gujarat (*iv*) Kachchha. Out of the total water resources of the state (Ground as well as surface water) the areas of Kachchh, North Gujarat and Saurashtara have 2 per cent, 11 per cent and 16 per cent water resources respectively which covers 58 per cent area of the state. These areas are subjected to draught having once in a five year frequency. These areas are one of the most water stressed area of the state as well as country. The mass migration tendencies seasonal as well as permanent towards Central and South Gujarat are observed regularly due to shortage of water, food and fodder. The main canal of Narmada dam has become new source of water for this region. The Sardar Sarovar canal based drinking water pipeline project is perhaps the biggest drinking water project in the world. This project will cover Saurashtra, Kachchh as well as parts of North Gujarat

Narmada River and Narmada Basin

The Narmada, the largest Westward flowing river, rises near Amarkantak range of mountains in Madhya Pradesh. It is the fifth largest river in the country and the largest one in Gujarat. It traverses Madhya Pradesh, Maharashtra and Gujarat and meets the Gulf of Cambay. The total length of the river from source to sea is 1312 kilometers (815 miles) while the length up to dam site is 1163 kilometers. (723 miles). The width of the river channel at dam site during high floods is 488 meter (1600 feet) and that during summer is 45.70 meter. (150 feet). The maximum recorded flood on 7th September 1994 was 70,847 cusecs (2.5 million cusecs) while minimum recorded flow in summer was 8.5 cusecs (300 cusecs.) The dam is designed for 87,000 cusecs (3.07 million cusecs) flood. The total basin area of the river is 97,410 square kilometer comprising 85,858 square kilometer in Madhya Pradesh, 1658 square kilometer in Maharashtra and 9894 square kilometer in Gujarat. The drainage area up to dam site is 88,000 square kilometer (Bhattacharya, 2007).

Sardar Sarovar Project

The Sardar Sarovar Project is one of the **largest water resources project** of India covering four major states–Maharashtra, Madhya Pradesh, Gujrat and Rajasthan. Dam's spillway discharging capacity (30.7 lakhs cusecs) would be third highest in the world. With 1133 cumecs (40000 cusecs) capacity at the head regulator, and 532 km. length, the Narmada Main Canal would be the **largest irrigation canal** in the world (Bhattacharya, 2007).

A concrete gravity dam, 1210 meters in length and with a maximum height of 163 meters above the deepest foundation level, is under construction across river Narmada. The dam will be the third highest concrete dam (163 meters) in India. In terms of the volume of concrete involved for gravity dams, this dam will be ranking as the second largest in the world with an aggregate volume of 6.82 million cu.m. The first is Grand Coule Dam in USA with a total volume of 8.0 million cu.m. This dam with its spillway discharging capacity of 87,000 cumecs (30.70 lac), will be the third in the world, Gazenba (1.13 lac cumecs) in China and Tucurri (1.0 lac cumecs) in Brazil being the first two. The gross storage capacity of the reservoir is 0.95 M. ha.m. (7.7 MAF) while live storage capacity is 0.58 M.ha.m. (4.75 MAF). The dead storage capacity below minimum draw down level is 0.37 M. ha. m. (2.97 MAF). The construction of dam up to 121.6 meter is over and construction of 456 km long main canal is almost completed (Bhattacharya, 2007).

The reservoir would occupy an area of 37,000 ha. and would have a linear stretch of 214 kilometer of water and an average width of 1.77 kilometer. The Full Reservoir Level (FRL) of the Sardar Sarovar Dam is fixed at RL 138.68 meters (455 feet). The maximum water level is 140.21 meters (460 feet.) while minimum draw down level is 110.64 meters (363 feet.). The normal tail water level is 25.91 meters (85 feet.) (Bhattacharya, 2007).

Sardar Sarovar Canal Based Drinking Water Supply Project

The Sardar Sarovar canal based drinking water pipeline project is perhaps the biggest drinking water project in the world. This project will cover Saurashtra, Kachchh as well as parts of North Gujarat (Ahmedabad, Mehsana, Banaskantha, Sabarkantha, and Panchmahal).The Figure 17.2 shows the map of Gujarat depicting various areas of the state. The capital cost of the project has been estimated to be Rs. 103.7 billion (2010 prices), which includes investments in trunk mains, water treatment plants, pumping machinery, civil works and water distribution. The project includes 48 bulk schemes, 123 distribution groups and a 3,250 km long pipeline network. The project is extremely important for Gujarat, as it will cover 53.3 per cent villages and 54 per cent urban centers of the state. It will cover 100 per cent villages in Ahmedabad, Amreli, Bhavnagar, Gandhinagar, Jamnagar, Junagadh, Kachchh, Porbandar and Surendranagar districts (Table 17.1). Similarly all towns are proposed to be covered in Ahmedabad, Amreli, Bhavnagar, Jamnagar, Junagadh, Kachchh, Patan, Porbandar, Rajkot and Surendranagar. In this project, water will be transported through bulk water transmission pipelines and distribution network to the doorsteps of villages and towns, which do not include in village/ town drinking water supply infrastructure development. Thereafter, in rural areas Panchayats will take over the responsibility of distribution, management, operation and maintenance of in village water supply systems whereas local urban bodies will be responsible for distribution management, operation and maintenance in towns.

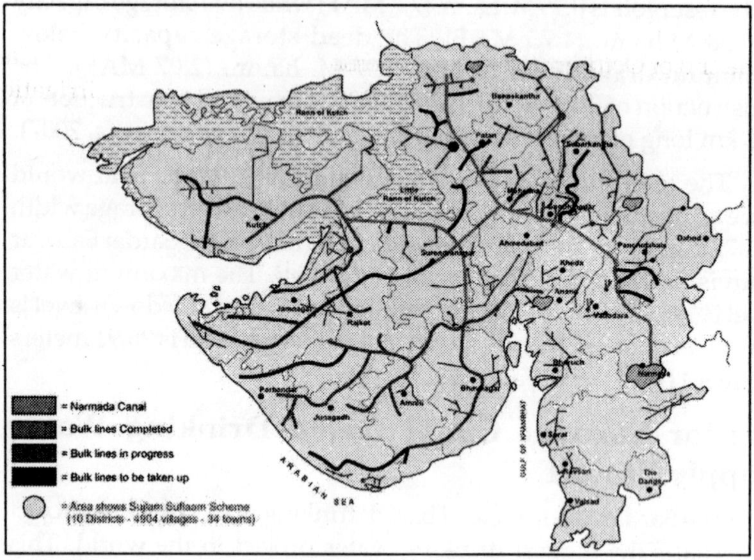

Figure 17.1: Bulk Lines Carrying Narmada Canal Water (*Source*: nca.gov.in)

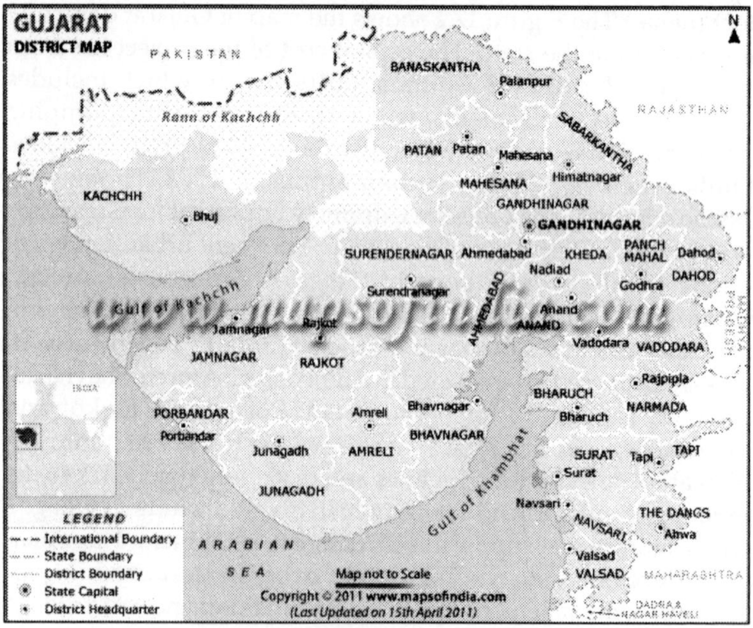

Figure 17.2: Map of Gujarat State

Table 17.1: District-wise Coverage of Villages/Towns in Revised Master Plan

Sl.No.	Districts	Number of Villages in the District	Villages Covered in the Master Plan		Number of Towns in the District	Towns Covered in the Master Plan	
			Number of Villages	Percentage of Villages		Number of Towns	Percentage of Towns
1.	Ahmedabad	546	546	100	25	25	100
2.	Amreli	615	615	100	8	8	100
3.	Anand	350	0	0	12	0	0
4.	Banaskantha	1244	816	65.6	6	2	33.3
5.	Bharuch	657	0	0	8	0	0
6.	Bhavnagar	790	790	100	13	13	100
7.	Dahod	693	0	0	4	0	0
8.	Dangs	311	0	0	0	0	0
9.	Gandhinagar	291	291	100	10	4	40
10.	Jamnagar	698	698	100	16	16	100
11.	Junagadh	923	923	100	12	12	100
12.	Kachchh	886	886	100	8	8	100
13.	Kheda	612	65	10.6	8	0	0
14.	Mehsana	593	464	78.2	8	7	87.5
15.	Narmada	552	78	14.1	3	0	0
16.	Navsari	374	0	0	9	0	0
17.	Panchmahal	1201	408	34	6	3	50
18.	Patan	517	421	81.4	5	5	100
19.	Porbander	182	182	100	5	5	100
20.	Rajkot	844	844	100	10	10	100
21.	Sabarkantha	1372	956	69.7	8	6	75
22.	Surat	722	0	0	15	0	0
23.	Tapi	445	0	0	3	0	0
24.	Surendranagar	650	650	100	7	7	100
25.	Vadodara	1548	0	0	16	0	0
26.	Valsad	450	0	0	17	0	0
	Total	**18066**	**9633**	**53.3**	**242**	**131**	**54.1**

Present Status

Though during its initial phase the project was quite slow in terms of implementation, by the end of October 2010 it has covered 6,513 villages, *i.e.* about 68 per cent villages against proposed coverage, and 114 towns *i.e.* about 87 per cent against the proposed coverage (Table 17.2). This is a commendable achievement. A 2,170 km pipeline has been laid (which is about 67 per cent of the proposed 3,250 km pipeline).

Table 17.2: Actual Coverage of Towns and Villages till October 2010

District	*1**	*2*	*3*	*4*	*5*	*6*
Ahmedabad	546	508	93	25	25	100
Amreli	615	456	74.1	8	8	100
Banaskantha	816	454	56.6	2	2	100
Bhavnagar	790	790	100	13	13	100
Gandhinagar	291	291	100	4	4	100
Jamnagar	698	668	95.7	16	16	100
Junagadh	923	360	39	12	9	75
Kachchh	886	743	83.9	8	8	100
Kheda	65	65	100	0		
Mehsana	464	464	100	7	7	100
Narmada	78	31	39.7	0		
Panchmahal	408	69	16.9	3	2	66.7
Patan	421	348	82.7	5	5	100
Porbandar	182	11	6	5	0	0
Rajkot	844	541	64.1	10	6	60
Sabarkantha	956	271	28.3	6	3	50
Surendranagar	650	443	68.2	7	6	85.7
Total	**9633**	**6513**	**67.6**	**131**	**114**	**87**

*1: Number of villages to be covered under Narmada water supply.

2: Number of villages already covered.

3. Percentage of villages already covered.

4. Number of Towns to be covered under Narmada water supply.

5. Number of Towns already covered.

6. Percentage of Towns covered under Narmada water supply.

Issues and Measures

To make Sardar Sarovar based drinking water supply project more sustainable it has to cope up with challenges like reliability of supply, adequacy of supply, un equal inter village distribution, un equal intra village distribution, in efficiency in operation and maintenance, water charges, recovery of charges, neglected local source of water, lack of systematic disposal of used water, village level management and neglected local water sources.

Conclusion

The long distance pipeline scheme under which water is transported through pipelines have several built-in weaknesses, which arise from the project concept and design on the one hand and the socio-cultural and administrative capabilities in the state on the other. However, the water stress areas of Saurashtra, Kachchh and North Gujarat are converted into water blessed areas due to Narmada main canal based drinking water supply project. The men and live stock have relived from the frequent shortage of water due to Narmada main canal water.

References

Bhattacharya, A., 2007. Report on "*Drinking Water Supply Schemes*", Department of Drinking Water Supply, Ministry of rural Development, India.

Gujarat Water Supply Sewerage Board, 2010. *Narmada Canal Based Projects, Progress Report.*

Hirway, Subhrangsu Goswami, 2008. *Dynamics of Crisis in Drinking Water in Coastal Gujarat with Special Reference to Saurashtra Coast*, Working Sample paper,CFDA,Ahemedabad

Mehta, Meera and Mehta, Dinesh, 2011. Urban drinking water security and sustainability in Gujarat, India. In: *Sardar Sarovar Project on the Narmada, Vol. 3.* Concept Publishing Company Pvt. Ltd., New Delhi.

Narmada Control Authority,*www.nca.gov.in*

Parthsarathy, P. and Dholakiya, Ravindra, 2011. Sardar Sarovar Project: Performance so far, plans and opportunities, In: *Sardar Sarovar Project on the Narmada, Vol. 3.* Concept Publishing Company Pvt. Ltd., New Delhi.

Parthsarathy, P. and Dholakiya, Ravindra, 2011. Sardar Sarovar Project: Background and overview. In: *Sardar Sarovar Project on the Narmada, Vol. 3.* Concept Publishing Company Pvt. Ltd., New Delhi.

Shah, T., Krishnan, S., Hemant, P., Verma, S., Chandra, A. and Sudhir, C., 2010. A case for pipelining water distribution in the Narmada Irrigation System in Gujarat, India. Colombo, Sri Lanka: International Water Management Institute, 25p.

SSNL, www.**sardarsarovardam**.org.

Talati, J. and Liebrand, J., 2003. Evolving institutions for irrigation management in Sardar Sarovar Project command, IWMITata Water Policy Program, Anand, India.

Vyas, Jay Narayan, 2011. Sardar Sarovar Project: The engineering marvel. In: *Sardar Sarovar Project on the Narmada, Vol. 1.* Concept Publishing Company Pvt. Ltd., New Delhi.

Chapter 18

Bringing Relevance to Agriculture in Hills in the Face of Emerging Natural and Human Challenges

☆ *Sanjeev Vasudeva and Punam Sharma*

Given the emerging population : productive-land ratio, and declining interest in agriculture given its reduced economic viability and depleting interest of youth, the problem of food security will not wane until some serious steps are undertaken to address issues that face us today. Within this overall scenario are included issues of centralized influences, be it the state sponsored Public Distribution System or aspects of Globalisation at a larger scale. As if these challenges were not enough, aspects of climate change and governance add to the melee of challenges that deserve serious thinking. Finally geographical regions as inherited by hill states further add to the complexity.

There is also a need to reflect on what agriculture is pursued for? Most efforts tend to isolate the activity from needs of food security that too needs to draw its bearings from needs of human nutrition. Simply put, higher productivity in agriculture need not mean that a nation can be termed as food secure, and furthermore being food secure need not point to the population being healthy from a nutritional intake perspective. A classic example that can be cited in this regard is the situation in Kinnaur and Lahaul-Spiti where it has been noticed that traditional practices followed by poorer sections of society have helped them rear healthier children in comparison to those of better off families, dependent on contemporary food approaches. These practices have a strong link to agriculture and many a times on an exclusive basis, meaning thereby that certain foods found in these regions are not easy to grow within the larger food production system. All of us involved with the subject of agriculture may like to stop for a moment and think what these linkages mean to our pursuit, and reorient our efforts as needed.

The complexity of food challenges being faced has thinkers baffled, largely reflecting through realignment of research and extension orientation the world over. We in our country lack the motivation to yet understand that these complexities will impact us more than other countries given our population pressure and resistance to change. The former is a well known phenomenon and the latter draws its qualifications from our hangover of the green revolution days that has influenced central thinking even for regions and areas that did not form part of the mid-60's initiated thrust. Areas that have been beneficiaries of the green revolution are needing to think more seriously about the cascading impact of their involvement, while those not having benefited on account of initial exclusion may feel more fortunate given the state of agriculture in regions where unsustainable intensification efforts have been pursued.

Hill states face unique challenges in terms of remoteness and their special geographical and food needs. They are also made to bear the skewed burden of upholding the cause of the environment translated exclusively in terms of keeping catchment intact by increasing area under forests; as if they are solely responsible to compensate for the problems created by all. Notwithstanding historical evolution of people- environment relationships, ecological elements in these regions have to undergo severe scrutiny and are kept under a more severe watch compared to other inhabited regions.

Aspects of remoteness and small land holding combined with climatic aberrations also make the activity of agriculture difficult to pursue from an economic viability point of view. On account of declining returns, especially from traditional crops, in recent times the increased phenomenon of migration and resultant shortage of working hands is leading to a reduction in the net sown area[1] that is a cause of worry. Reviewing this data against the total cropped area, one can observe that the comparative decrease in total cropped area is less than decrease in net sown area (1990-2005)[2]. This indicates intensification on cropped lands but reduction in net cropped area. All this in a region where land is scarce and its needs to serve humanity for food bears much importance.

An analysis of area under different crops indicates an increase in wheat and maize whereas the area share of other cereals, particularly traditional food based crops that include ragi and other common and small millets, declined from 5.9 per cent in 1972-73 to 1.3 per cent in 2004-05, with area under pulses getting more than halved during the same period. This increase in wheat and maize production needs to be viewed against a drop in production of pulses and the practice of mixed cropping, largely on account of higher application of nitrogenous fertilizers to meet needs of wheat and maize. During the same timescale, oilseed production area has declined from 2.58 per cent (1972-73) to 1.61 per cent (2004-05) and such a skew in agriculture can not be viewed as being favourable to the cause of sustainability.

The graph below presents the value added trend over the 7 year period between 1999-2000 to 2006-07 with horticulture enjoying high overall growth of roughly 166 per cent, livestock a steady growth of 54 per cent, and crop based agriculture experiencing see-sawing fortune to eventually end with a dismal overall growth of 13 per cent. These figures point to an annualized growth of 13.7 per cent for Horticulture, 6.7 per cent for Livestock, and 1.04 per cent for

Crops, with total agriculture growth pegged at 7.18 per cent, thanks to horticulture's contribution to overall kitty.

With declining economic gain promised by core agricultural activity, it is not surprising to find most farmers turning to horticultural activity that needs less labour and promises higher returns; though not bereft of its own set of problems. Among other

---

1 From 583 thousand hectare in 1990-91 to 542.7 thousand hectare in 2004-05.

2 NSA declined by 40.3 thousand hectare whereas total cropped area declined by 30.4 thousand hectare.

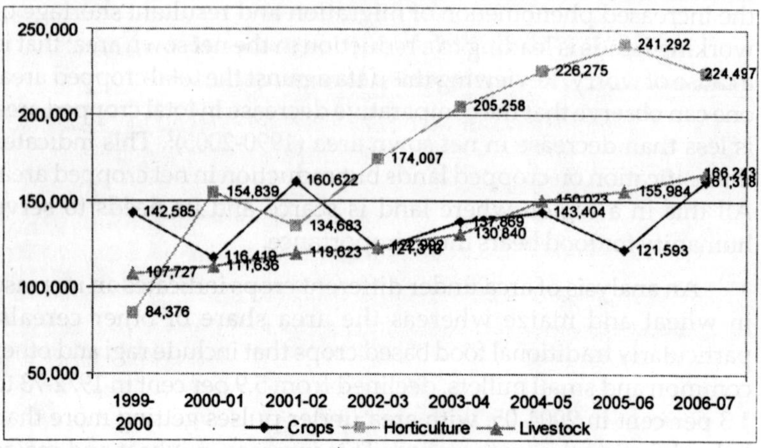

Figure 18.1: Gross Value Added Over Years 1999-2000 to 2006-07 (Rs. Lakh)

Source: State Agriculture Plan, Volume XII

reasons attributed to such a decision being taken is the unsustainability of agriculture pursuit given its poor record of impact with relation to soil and water, thus pointing to perennials as a way forward. Populist needs lead to policy support getting mustered and this too has shown its head in state formulated policies of late. Little can be seen on the horizon that points to a changing situation in days ahead, and if this indeed continues in the currently manifested form, it can have serious repercussions for such regions; not only from a food security point of view but also livelihoods. This paper attempts to present such issues and suggest a possible way forward that hopefully will serve as an effective thinking point for policy makers.

Himachal Scenario

Land Use

Land is scarce and this needs to be borne in mind while recognizing that agricultural activity caters to livelihood activity for 68.48 per cent of Himachal citizens, and contributing 30 per cent to the state's gross domestic product. To briefly set the backdrop for the region being discussed, the following graph gives an indication of the features of land use:

Extract from the State Agriculture Plan, Volume XII report

The growth trends of the NSDP in the state indicate that except for the second half of 1980s, agriculture sector fared poorly throughout the decades of 1980s and 1990s (Table 2.9). It is only during the period 2000-01 to 2005-06 that this sector registered a significant growth of 6.97 per cent.

Furthermore, a decomposition of the gross value of output in this sector revealed that horticulture sector is emerging as a prime mover of the growth in agriculture in the last few years. This is amply borne out by the fact that its share in gross value of output has gone up from 25.21 per cent in 1999-

2000 to 40.67 per cent in 2006-07 (Table 2.10 and Figure 2.4). The share of crop production, on the other hand, has come down from 42.60 to 29.22 per cent during the same period. The contribution of livestock sector has remained almost stagnant at 30.0 per cent level. The increasing contribution of horticulture in the state also becomes evident from the fact that the gross value added (GVA) in this sub sector increased from Rs.843.76 crore in 1999-2000 to Rs.2244.97 crore in 2006-07 registering a highly significant growth rate of 13.7 per cent per annum.

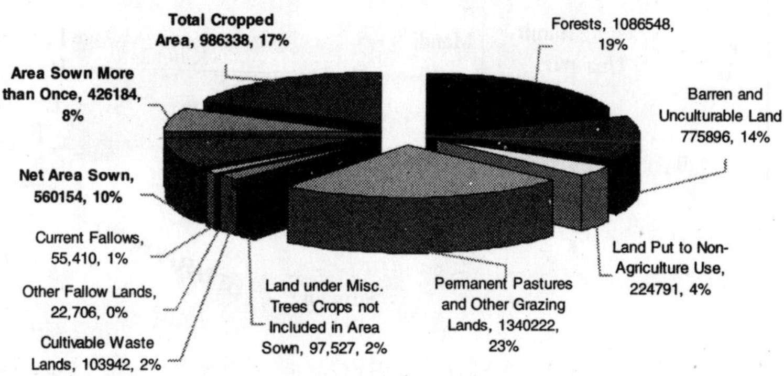

Figure 18.2: Land Utilisation Pattern of Himachal

Source: Himachal Pradesh Agriculture Department Website

Only 18-20 per cent of the area is irrigated with the rest being rainfed that is a good indicator to the adaptive needs that need to be applied on an ongoing basis given the over dependence on natural factors by farmers, most of whom have small land holdings.

Agro-ecological Zones

Himachal is broadly divided into 4 agro-ecological zones as below:

Shivalik Hill Zone

With sub tropical climatic features, this zone consists of foothills and valley area from 350 to 650 meters above mean sea level, It occupies about 35 per cent of the geographical area and about 40 per

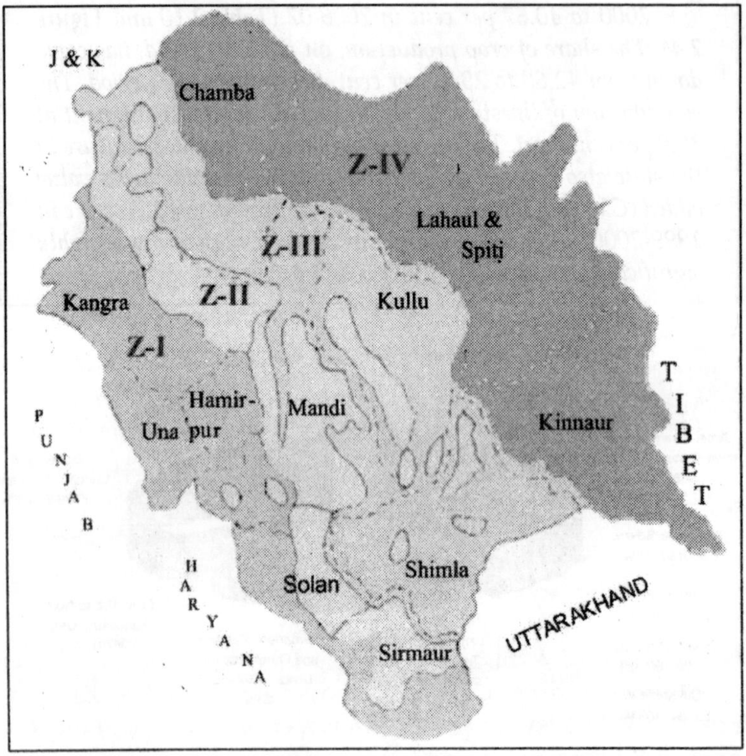

Figure 18.3: Agro-ecological Regions of Himachal

Source: State Agriculture Plan, Volume XII

cent of the cultivated area of the State. The Zone has problem of severe soil erosion. The major crops grown in this Zone are Wheat, Maize, Paddy, Gram, Sugarcane, Mustard, Potato, Vegetables etc.

Mid Hill Zone

This zone extends from 651 meters to 1,800 meters above mean sea level. Having a mild temperate climate, it occupies about 32 per cent of the total geographical area and about 37 per cent of the cultivated area of the State. In addition to the problem of soil erosion, the zone's soils are deficient in nitrogen and phosphorous and have poor water and nutrient holding capacity. Major crops grown are Wheat, Maize, Barley, Black Gram, Beans, Paddy etc. This zone has very good potential for cultivation of cash crops like off-season Vegetables, Ginger, and production of quality seeds of temperate vegetables like cauliflower and root crops.

High Hill Zone

This zone possesses good pastures and meadows and lies between 1,801 and 2,200 meters above sea level with humid temperate climate and alpine pastures. It covers about 35 per cent of the geographical areas and about 21 per cent of the cultivated area of the State. Commonly grown crops in this region are Wheat, Barley, Lesser Millets, Pseudo-cereals (Buckwheat and Amaranthus), Maize and Potato etc. The area is ideally suited for production of quality seed Potato and temperate vegetables.

Cold Dry Zone

It consists of Lahaul-Spiti and Kinnaur Districts and Pangi Tehsil of Chamba District that lies about 2,200 meters above mean sea level and occupies roughly 8 per cent of geographical and 2 per cent of total cultivated area of the State. Major crops grown in these upper reaches are Wheat, Barley, Pseudo-cereals like Buck wheat and Amaranthus. This region too is well suited for production of temperate and exotic vegetables and their seeds, Seed Potato, and Peas as green and for seed purposes.

As can be observed, there are differing needs of agriculture in each of these regions and to that extent differing strategies need to emerge to meet needs of the people residing there. While the state department has been providing for this through effective measures, there is a need to look at the emerging situation from a more holistic perspective and we do hope this paper can evoke such a discussion in days ahead.

Land Holding

In terms of land holdings, Table 18.1 shows that nearly 2/3rd farmers hold less than 1 hectare of land. This impacts agriculture in many ways, (i) Pressure to adopt means of intensification leading to unsustainable practices, (ii) Unviable pursuit from economy of scale standpoint, (iii) inverted land to man ratio from food supply point of view, (iv) declining interest to address the problem at hand and make corresponding investment.

Given this situation, it is of utmost importance to find solutions that address the emerging situation, in this case make pursuit of agriculture feasible and viable with improved dependence on local means. At the same time land aggregation principles need to be applied to make pursuit of agriculture meaningful.

Labour

The proportion of agricultural labourers standing at 3.15 per cent considering the large number of population (65.33 per cent) involved with agriculture as cultivators is an indication of the self employability feature of this sector. Adding both these numbers points to more than 2/3rd of the people being dependent on agriculture. Among the districts, the dependence on agriculture is highest for Kulu (78.63 per cent), Sirmaur (74.09 per cent), Mandi (73.95 per cent) and Chamba (73.44 per cent) thus pointing to the need to bring in measures leading to improved chances for those involved with agriculture in these regions.

Thrust Areas Identified

On the basis of given factual conditions, the Himachal government has identified the following thrust areas for agriculture:

1. Diversification of area from traditional crops to commercial crops where irrigation potential has been created. Farmers shall be motivated to produce organic vegetables without the use of pesticides and chemical fertilizers.
2. Development of rainfed areas through watershed approach on a large scale for efficient use of natural resources. Increased funding shall be arranged under RIDF.

Table 18.1: Distribution of Land Holdings in Himachal Pradesh, 2001

Size of Holding (ha)	*Category*	*No. of Holdings (lakh)*	*Area (lakh ha.)*	*Average Size (ha.)*
Below 1.0	Marginal	6.15 (67.3 per cent)	2.52 (25.8 per cent)	0.4
1.0 - 2.0	Small	1.74 (19.1 per cent)	2.45 (25.0 per cent)	1.4
2.0 - 4.0	Semi-medium	0.90 (9.8 per cent)	2.43 (24.8 per cent)	2.7
4.0 - 10.0	Medium	0.31 (3.4 per cent)	1.76 (18.0 per cent)	5.7
10.0 and above	Large	0.04 (0.4 per cent)	0.63 (6.4 per cent)	15.7
Total		9.14	9.79	1.1

Note: Figures in parentheses are percentages of total.

Source: Agricultural Census, 2001, Directorate of Land Records, Government of Himachal Pradesh.

Table 18.2: Composition of Total Workers in Himachal Pradesh

Districts	*Total Workers*	*Cultivators*		*Agri. Labourers*		*Household Industry*		*Other Workers*	
		Number	*Per cent*	*Number*	*Per cent*	*Number*	*Per cent*	*Number*	*Per cent*
Bilaspur	166,708	114,244	68.53	2,923	1.75	2,727	1.64	46,814	28.08
Chamba	230,452	167,590	72.72	1,655	0.72	2,780	1.21	58,427	25.35
Hamirpur	205,405	143,535	69.88	3,291	1.6	2,916	1.42	55,663	27.1
Kangra	588,994	335,357	56.94	39,534	6.71	19,345	3.28	194,758	33.07
Kinnaur	47,811	30,977	64.79	1,101	2.3	871	1.82	14,862	31.09
Kullu	216,513	164,646	76.04	5,590	2.59	2,733	1.26	43,544	20.11
Lahaul & Spiti	21,088	11,160	52.92	339	1.61	136	0.64	9,453	44.83
Mandi	454,292	329,472	72.52	6,518	1.43	6,747	1.49	111,555	24.56
Shimla	370,223	237,490	64.15	9,814	2.65	4,392	1.18	118,527	32.02
Sirmaur	225,872	161,598	71.54	5,767	2.55	2,595	1.16	55,912	24.75
Solan	263,445	143,630	54.52	6,309	2.39	3,291	1.25	110,215	41.84
Una	201,658	115,171	57.11	11,330	5.62	3,986	1.98	71,171	35.29
HP	**2,992,461**	**1,954,870**	**65.33**	**94,171**	**3.15**	**52,519**	**1.75**	**890,901**	**29.77**

Note: Percentages are with respect to total workers.

Source: Census, 2001, HP.

3. Rainwater harvesting is another area that will not only provide life saving irrigation to crops but shall also recharge ground water and check erosion. The department shall seek financial assistance from Govt. of India for small irrigation tanks/shallow wells and pumping set.
4. Increase in maize productivity through high yielding hybrids.
5. Adoption of precision farming practices (poly-houses and micro-irrigation).
6. Organic farming shall be a thrust area.
7. Post harvest management and efficient marketing system.
8. Farm mechanisation with special reference to hill agriculture shall be given major thrust in the years to come. This is necessary to reduce cost of cultivation in view of high cost of labour.
9. A strong research-extension interface directed towards problems oriented research programmes. Research projects to be identified and funded in problem areas.
10. Extension reforms through public-private partnership.
11. Agro-processing and value addition.
12. Increase in productivity and quality.
13. Application of Biotechnology in the field of Agriculture shall be explored.
14. Soil Testing and Soil Health Cards will be given an impetus.

As can be seen, the government recognises the need to give a boost to agriculture and has included concerns of environmental degradation as covered through soil and water conservation measures. Further and more importantly, under proposed interventions as part of the State Agriculture Plan, Volume XII document one finds many interventions given below that indicates the positive spirit proposed to be enthused by the H.P. Govt. in its agriculture related efforts, all supported by Conservation Agriculture:

- ☆ "Improvement of physical, chemical and biological parameters on the basis of soil health cards proposed to be issued to all farmers." (Pg. xix)

- "Promotion of mechanization conducive to hill farming....to reduce labour and to provide relief to the women folk" (Pg xix)
- "*Ex-situ* and *in situ* conservation of minor millets cereals" (Pg xx)
- "In situ soil and water conservation....employing different measures" (Pg xx)
- "Improvement of support lands (private grasslands) by employing different measures for the control of invasive shrubs, weeds, etc." (Pg xx)

Under Researchable issues one finds the following:

- "Evaluation of resource conservation technologies (like conservation tillage, deficit water management, pressurized irrigation systems, nutrient-water interaction studies, recycling of waste organic residues, etc) for irrigated and rainfed areas." (Pg xxvii)

Under the threat part of the SWOT Analysis of the state one finds

- "Nearly four-fifths of cultivated area suffers from high to very high soil erosion in state." "The global warming induced climatic changes such as irregular rainfall pattern, unusual seasonal variations, early snow melting, deglacialization, etc. have resulted in many crop management problems in the form of altitudinal shifts in the cropping pattern, water management issues, new diseases and pests. This has started impinging upon the sustainability of agricultural production systems in the hills."
- "The commercialization of agriculture... excessive and imbalanced use of chemicals.....resultant pollution [is manifested in] disappearance of many small millets, some pulses and oilseeds" (Pg 44)
- "Many obnoxious weeds...have engulfed the cultivated land...resulting in the declining productivity of these lands on the one hand and escalating the cost of cultivation on the other" (Pg 44)

The above mention description shows the recognition of climate change, loss of soil and soil nutrition depletion, problems of weeds and new disease, loss of crop diversity, i.e. disappearance of small millets and the need to research on resource conservation technologies. Yet, if one refers to section Sectoral Outlays and Yearly Allocations, activities mentioned under Natural Resource Conservation and Management continue to be soil conservation through engineering measures, water harvesting and land improvement. This in the face of the section on Proposed Intervention (Pg. xix) explicitly mentioning the need for improvement of physical, chemical and biological parameters of soils as one of the proposed interventions. This shows that resource conservation technologies while visualised as being of critical importance have not gone beyond being looked at as only a research topic at this stage.

Emerging Issues and Influences

Productivity Gap

If one looks at the crop mix that constitutes agriculture in the hills, we are faced with two kinds of situations, (i) crops that are shared in terms of growth in the less agriculturally challenging regions of the mainland, and (ii) those that are in a manner of speaking exclusive to hill regions. It also goes without saying that the former is easy to procure while the latter has to be relied upon for its growth in the local context, and to an extent, the state bears responsibility to address this need. Obviously, over time this has started to influence the pattern of food consumption and hence horticulture production in the hill region, possibly eating into the area that was earlier devoted to growing crops exclusive to the region.

Considering the scope of this paper, without going into too much into detail on aspects of nutrition, its impacts thereof, and need to support traditional cropping systems; even if we try to look at comparative figures of agricultural productivity for crops grown in both regions in table below, evidence builds up on the need to address issues faced. We may like to remind ourselves that with relation to the subject of food, underestimating and thus not positioning needed long term efforts can have serious consequences. As such, a "business as usual" approach may not be the right way to address issues. Further, agricultural productivity enhancement needs to be addressed from a local sustainability perspective and the role of policy is very important to encourage achievement of such a goal.

Crop	*National Productivity (per ha.)*	*State Productivity (per ha.)*	*H.P. Productivity as per cent of National*
Rice	1,907 kg	1,364 kg	71.53
Wheat	2,577 kg	1,426 kg	55.34
Corn	1,714 kg	2,078 kg	121.24
Pulses	609 kg	701 kg	115.11
Oilseeds	896 kg	439 kg	71.53

The authors believe that this issue needs to undergo a correction and if past efforts have been seen not to bear fruit, newer approaches have to be given a chance and in this regard Conservation Agriculture may be one such approach deserving consideration. A more detailed discussion on this concept follows later.

Need for Sustainability: Human, Ecological, and Financial

Agriculture for its sheer scale of people involvement deserves to be addressed more seriously. On account of this very nature, issues of sustainability too need to become effective cornerstones in its pursuit. What needs to be borne in mind though is the fact that the more exotic the approaches advocated, the more isolated will be its uptake and it is thus very important that the solution advocated needs to be simple to pursue and make eminent sense to the farmer to encourage its wide adoption and mainstreaming.

In this regard it may be worthwhile to review a scenario presented in the context of Lahaul-Spiti. "Sustainable agriculture is highly difficult in mountain areas because of fast decreasing soil fertility, high degree of fragile topography and consistent soil and water erosion problems. The frequent land slides in the fragile eco-system often lead to fast depletion existing forest cover. The cold desert in Himachal Pradesh constitutes 42 per cent of the total geographical areas. These areas suffer from acute shortage of fuel wood, fodders apart from soil and water erosion problems. Even though many engineering structures have been raised to protect these fragile eco-system, but all in vain. In these extreme climatic conditions, environmental protection and conservation are Himalayan tasks. Hence the major emphasis and obvious solution is only massive afforestation programme through a soil binding and water conserving plant and that should not only be the long term environmental and resources conservation mechanism but also the best sustainable short term strategy for the mountain people."

Source: http://hplahaulspiti.nic.in/InnovationDDP.htm

Regardless of the single point advisory note it ends on, the fact remains that there is an accepted understanding to address needs of soil and water conservation to ensure continued utilisation of land for mankind's needs. It is in this context that the largest contribution can be made by Conservation Agriculture helping restore the state of lands and livelihoods of the hill farmer. This form of agriculture makes sense since it works at addressing needs of sustainability not only in natural terms (improved soil and water conservation, and reduced vulnerability) but also in relation to human labour (no tillage reduces aspect of human effort), and financial (reduced cost from lack of tillage and reduced input cost on account of improved soil health in years ahead).

Integrative Efforts Needed

The hills have pasture lands that give opportunities to even poorer households to benefit from livestock rearing. Dairy farming goes beyond being a mere livelihood pursuit and in fact represents a traditional culture that most feel the need to protect and foster. Given this backdrop, it is very easy for hill regions to look at agriculture as an integrative activity bringing together needs and outputs of farming and dairying for mutual good. Conservation Agriculture works on these very principles and encourages crop-livestock interaction considering its needs for crop residue to deploy on lands as mulch. Recognising that there are competitive pressures on crop residue emerging from needs of domestic fuel and fodder, livestock rearing is included as an area to address to free sufficient biomass needed to serve as mulch. On another count, looking to needs of biomass, an afforestation program or a pasture development program would be found to be in order to make Conservation Agriculture sustainable in the long term.

Economic Considerations-vs-Human

As discussed earlier, it is not surprising to find interest shifting to more economically viable options. Feasibility of such crops has improved over time given increased access through investments in rural infrastructure and opening of markets, many such efforts being successful in developing reverse linkages to the farm gate. The case of apples would be most worthy to cite. Over the last decade, effective linkages have made apples affordable to eat in distant parts of the country. This diversification move has worked for the people of Himachal Pradesh. However such horticulture crops are highly dependent on market forces and market swings. It will not be surprising to hear of many other stories of apple farmers who have had to bear the brunt of such events with rotting stock held in their hands and desperate sales being made. Similar stories may emerge from off-season vegetables and organic farming once volumes pick up. In any case this shift in interest and priority works to benefit other players along the value chain much more than the farmers who go through the trouble of diversified farming. There is thus a need to understand such situations and balance high value pursuit with staple food crop needs.

While one can not have everything and there is no one clear line of action to be pursued, we need to understand and rationalise our actions in terms what such a shift means to satisfy the needs of the smallholder farmer. Other than meet needs of nutrition grown in one's farmland, the culture is now shifting to growing cash crops and buying food. This subject is beyond the scope of this paper but it will suffice to state that policy direction given by the state can be a strong influencer of such activity shifts. The government in all its wisdom should encourage growth of traditional crops and help them find their place in the new world that can be made to happen in a co-existence mode. Be it at the market end, or by building awareness and respect for such a practice, policy support will be the key driver to influence a farmer's decision. If not at an individual farmer's level, the government needs to internalise the larger need to attain food self-sufficiency for its people by reducing external dependence and associated vulnerabilities.

Given the high rate of pursuit of agriculture, what would be essential would be initiatives and support to help farmers pursue this activity sustainably, and more importantly reducing the risks and vulnerability associated with it. Whether these have to be faced in terms of natural resource degradation or on account of newly perceived threat of climate change, Conservation Agriculture efforts have a positive track record to share. In any case even for horticulture crops, Conservation Agriculture has known to accumulate benefits wherever mulch application has been practiced. Thus the practice is equally relevant to horticulture as to basic agriculture.

Natural Degradation Aspects of Vulnerability

Another reason stated for shift from agriculture to horticulture is the degradation of environment faced while pursuing agriculture, with its particular impact on soil erosion. While earlier practices and the knowledge built around such efforts may have warranted such a shift, modern day scientific knowledge indicate possible role reversals if Conservation Agriculture practices of no-till and mulch based farming are pursued. To put it briefly (detailed description is covered later), the former encourages minimal soil disturbance with the latter reinforcing this need and also contributing to needs of water conservation, temperature moderation, and improved soil health.

Commencing year 2000 there has been a decline in overall rainfall in Himachal Pradesh and this development is bound to affect pursuit of agriculture and calls for adaptive mechanisms to be positioned. Beyond this, agriculture is also blamed in the context of climate change for being a cause on one hand and a sufferer on the other hand. If one were to assume climatic uncertainty as another culprit, Conservation Agriculture on account of early sowing and harvesting to an extent does offer a way out of such a situation. Its practice of improving carbon formation within the soil helps reduce GHG emission to some degree. It has also been noticed that with no-till mulch-based farming the crops/plants build better resilience to natural phenomenon, thus helping the farmer by reducing his vulnerability. This reduced vulnerability, backed by reduction of soil erosion and restoration of soil health can form as strong pillars of growth to bring back interest in farming for a large section of the state's population.

Food Crops Form Essential Basis of Nutrition Delivery

The agriculture system is the base through which is reflected the state of any region's nutritional needs. Shifts in agriculture cropping and availability of food have of late, among other reasons, been responsible for declining nutritive state of poor people. As if not to be seen as a malady of the poor, very many better off households have reported dropping nutrition levels that are a bane to a nation's workforce being able to contribute effectively to tasks taken up. If within this melee, availability of food in quantitative (production) and qualitative (nutritive mix) terms is not addressed satisfactorily, delivering similar level of supplementary nutrition through an alternative route such as nutritional supplements will create a great burden on the exchequer. This needs to go beyond viewing it as a financial burden and must be further viewed in light of the ability to achieve the task effectively, knowing competency issues prevalent in the health sector or even if one is to view the PDS as an indicator. As such it is important for a state government to influence agriculture based on nutrition and food security needs of its population, rather than be influenced by the fate that an unplanned agricultural situation may bring upon us.

In light of this, it would be pragmatic for the state to influence cropping mix through policy support measures that encourages self reliance and self sufficiency at a local "production cluster" level. This advice is more relevant to a state such as Himachal as compared to others given its topography representing itself in terms of remoteness and difficulty of access. More than any other reason, in cases of disaster this strategy will go a long way to alleviate problems if ever such a situation is created.

Cluster Approach for Mutual Dependence & Improved Viability

The challenges posed by a hill region are not only physical in nature but can also challenge the best financial minds wishing to find solutions of viability. Going beyond this, there is a need to foster a spirit of inter-dependency for collective good. As such a cluster based mechanism needs to be positioned at the farm level to make efforts viable by avoiding duplication of efforts on the input side (nursery, agri-input sourcing, service delivery) or the output end (aggregation of produce to process or reach to the market).

While many initiatives may have been attempted in the past with varying results, such experiences will need to be learnt from to have the cluster system better deliver on expectations set from it.

While introducing Conservation Agriculture practices, aspects of awareness and capacity building form important pillars on which this transformational program will need to be launched. These will be best achieved through the collective mechanisms suggested. Beyond this, if confidence develops within this cluster system, it could be the beginning of a "land co-operative" system of agriculture engagement that could prove to be a boon for the hill farmers given the small parcels of land they have to contend with making the entire process of farming highly unproductive.

Policy – The Key Influencer

Policy can be the single biggest driver of actions on the ground that can also pretty much seal the fate of agriculture and farmers. Thus efforts need to be positioned to link needs of nutrition to food security and eventually to agriculture. Considering the varied nature of Himachal's geographical features, sufficient representation will need to be given to regions with special needs on the basis of which broad classifications will have to be drawn up based on identified socio-natural demands that will emerge. Broadly speaking, these will follow basic demarcation principles based on agro-ecological regions prevalent, fine tuning may be needed to accommodate needs of remote regions.

Such policy measures will need to broadly address the following points (covered in detail later):

- ✰ Awareness building and dissemination
- ✰ Field demonstration trials to establish results on "region x crop x season" basis
- ✰ Incentive to influence cropping
- ✰ Encouragement to cluster based farm development
- ✰ Scheme for easy availability of agri-inputs

What can Conservation Agriculture mean to Himachal Pradesh?

Basics about Conservation Agriculture

While environmental concerns related to agriculture are increasing, there is a need to see agriculture as part of the larger environmental concern, more relevant to regions that are ecologically sensitive. Conservation Agriculture (CA) through its resource conservation approach can be among the strategies that can point to a new and improved way forward as a mainstreamed agricultural system. With climate change becoming a reality there is increasing need to adopt mitigative measures for long term benefit while ensuring adaptive measures to meet current day needs. Sufficient evidence has accumulated to show that CA practices offer both these features. Given its ability to offer short term gains to farmers through savings on input costs, material, finance and labour, yet position relevant mechanisms related to long term goals such as climate change adaptation, CA can become a preferred practice.

Basic principles of CA include:

1. Minimum soil disturbance, *e.g.* through no-tillage
2. Keeping soil covered by maintaining crop residue and growing cover crops
3. Adopting diversified crops and sequences

CA helps to address climate change and provide a way to adapt to new situations by:

- ☆ Moderating the impact of high atmospheric temperatures on soil temperatures and plants by soils cover providing crop residues etc.
- ☆ Reducing the adverse impact of stress/drought periods or excess water periods through enhanced in-situ storage, reduced runoff and erosion, reduced water losses through evaporation
- ☆ Among CA, no-till is one of the few agricultural practices that can deliver services that benefit farmers, society, and the environment, including benefits such as reduced erosion, C sequestration, energy conservation, and decreased N loss.

Conservation agriculture can provide cheaper and more reliable supply of food while addressing environmental needs such as improved soil health, reduced water pollution and runoff, and decreased greenhouse gas emissions. There is increasing attention being given to farmers practicing CA, particularly those adopting reduced tillage systems that offer advantages of reduced costs, and potential to reduce carbon emissions. There is a need to understand

and learn from responses and strategies practiced by such farmers to arrive at an overall strategy, aimed at further assisting them to combat climate related impacts.

Conservation Agriculture as an Adaptive Strategy

Farming practice based on the basic principles of Conservation Agriculture (CA), when adopted in an integrated fashion over a period of time contribute to sustainable increases in crop productivity, improving soil health, biodiversity and in reversing processes contributing to land and water degradation. It has great potential to improve soil carbon sequestration and decrease net emissions of CO_2 and other greenhouse gases that contribute to global environmental security.

Soil organic carbon represents a key indicator for soil quality for both agricultural and environmental functions. Elements of CA offer opportunities to enhance carbon sequestration in soils and biomass. CA practices through change in agriculture technology and management offers both mitigation and adaptive strategies to combat climate change as indicated in Table 18.3.

While Conservation Agriculture principles are fairly universal in nature, technologies aimed at translating these into practical solutions have to be specific to regions and the existing farming system. The principles of CA are not new, they have been known and tested over time. What is new is the recognition of the need to pursue these in an integrated fashion – and this then is the real challenge. Development and promotion of agricultural practices and technologies have to be in the context of local farming situations, that will vary from region to region. It is here that the farmer's experience and knowledge can form a sound base and an input to develop new and improved practices leading to a knowledge base for improved adaptive capacity.

Why Conservation Agriculture Makes Sense to Himachal Pradesh

PACA's judgement based on its rainfed field experience (see box below) has established that CA is practical to pursue from a small farmer perspective and this is the most heartening aspect of this approach. Another very important aspect is the principle of environmental restoration that forms the underlying principle of the approach, something very important to the priorities of the state.

Table 18.3

Elements of CA	*CA's Contribution to Mitigate GHG Emission*	*CA's Contribution as an Adaptive Strategy*
Develop and promote management practices that are based on following key principles	Reduced CO_2 emission due to tillage induced oxidation of soil organic matter	Moderating the impact of high atmospheric temperatures on soil temperatures and plants by soils cover providing crop residues etc.
Minimum soil distribution, *e.g.* through no-tillage	Increased CO_2 sequestration through building soil organic matter/ increased biomass	Reducing the adverse impact of stress/drought periods or excess water periods through enhanced *in-situ* storage, reduced runoff and erosion, reduced water losses through evaporation.
Keeping soil covered, *e.g.* by maintaining crop residue, growing cover crops etc.	Reduced NO_2 emissions by enhancing the use efficiency of nitrogenous fertilizers	
Adopting diversified crops and sequencing	Reduced CH_4 emission through opportunities to better manage livestock and rice paddy	

PACA Field Trial Experience in Vidarbha

Over two seasons in 2010, trials were pursued by poor farmers that set the ground for an encouraging alternative approach that not only gave required yields but also cut down on cost and time. You can access the story of kharif sowing for cotton and soya crops covered in PACA newsletter from http://www.conserveagri.org/PACA per cent 20Newsletter per cent 20Issue per cent 2015.pdf.

The results of these trials led to further enthusiastic pursuit in rabi when wheat and gram were taken up with equally encouraging results. This story is covered in PACA newsletter no. 17 (pages 6–9) dowloadable from http://www.conserveagri.org/PACA per cent 20NL17.pdf. More heartening was the fact that beyond the 2-3 acres of wheat and gram planned by PACA for rabi, one of the farmers went ahead and sowed sorghum (jowar) on his own, driven by his new found confidence. Another lady in a distant village, who had heard about the approach from her husband (who was not convinced) went ahead and sowed half an acre of wheat on her own.

In a more direct manner, the following can be stated as reasons why Conservation Agriculture deserves a serious consideration with relevance to needs of Himachal Pradesh:

- ☆ Principles of Conservation Agriculture are easy to understand and internalise since they acknowledge traditional knowledge resting within farmers
- ☆ Since it encourages local self-dependence means, it becomes relatively easy to roll out once experience based conviction is built in the farming community. This means much to a region where remoteness is an underlying situation
- ☆ CA is in line with needs, practices and priorities of hill regions given its focus on soil and water conservation, integration, diversification, and ecological restoration
- ☆ With its ability to give short term gains through cost and time saving measures, it is able to achieve long term gains of ecological rebuilding by maintaining continued interest of farmers for a period of time. Later it helps reduce input costs given the improved natural state of farm lands it is practiced on
- ☆ It has equal relevance and important role to play in the context of horticulture pursuit and gives room to needs of livestock and biomass production
- ☆ With tillage activity done away to quite an extent, it becomes more gender friendly to pursue since man's involvement for tilling now stands reduced. However one must be careful to observe that this should not result in increased burden being put on women for agriculture, already burdened otherwise

Way Forward – Steps Needed to be Initiated

To begin, a fair chance needs to be given to this form of agriculture that could spell hope for the poor smallholder farmers of Himachal and this will need to translate into effective steps taken, possibly as suggested below:

1. Initiate a stakeholder consultation within different socio-agroecological situations leading to a state-level discussion on the subject. This consultation needs to dwell

upon and help arrive at an understanding of an issue such as "cash crop opportunity-vs-food self sufficiency in local terms"

2. Policy issues emerging will need to be acknowledged and a direction taken to debate and effect support mechanisms through schemes directed at encouraging the sustainable route pointing to improved agriculture practices in days ahead. This exercise will need to be viewed as an evolving exercise and not one of one-time nature. As indicated earlier, these will revolve around:
   - ☆ Awareness building and dissemination
   - ☆ Field demo trials to establish results on "region x crop x season" basis
   - ☆ Incentive to influence cropping
   - ☆ Encouragement to cluster based farm development
   - ☆ Scheme for easy availability of agri-inputs and machinery as identified
3. A low-level awareness program about Conservation Agriculture can be piloted and experiences gathered for larger adoption
4. This could be followed up by first sensitising department officials and then undertaking field trials by the Department of Agriculture in different zones identified as above
5. Experiences gathered need to be understood to fine tune larger roll out efforts if found deserving and package of practices could be developed for wider engagement across the state. We may mention that efforts will work to present CA as a mainstream change rather than a haloed approach limited to the few who understand it
6. At this stage CA's adoption in ongoing efforts such as watershed programs and departmental pursuits will need to be integrated within their Annual Work Plan & Budget effort
7. Capacity development programs relevant to needs of enablers and Himachal smallholder farmers will need to be developed and delivered

8. Wide scale adoption and monitoring of ongoing efforts with baseline and evaluative mechanisms would finally need to be effected for the benefit of Himachal farmers if all earlier efforts point to such an eventuality

Conclusion

All in all Conservation Agriculture promises to deliver on various counts of sustainability, much needed in the current day context where economic, social, and natural challenges are making availability of food an issue to contend with. CA offers hope considering its ability to address needs of climate change and resource degradation and its mix of short and long term benefits. In this manner it is able to do justice to meet current day needs of farmers while helping them address larger issues pertaining to the cause of environment and their own natural state of existence in days to come. In all fairness, what is proving itself across the world needs to be given a chance to be understood in the context of a hill situation for the good it promises.

Chapter 19

Design Principles for Community–Based Natural Resource Management: A Case of Community Managed Canal Irrigation System*

☆ *Shailendra Singh Bisht and Sachin Kumar*

ABSTRACT

It has been observed that concomitant changes in water quantity and quality due to climate change are likely to increase the vulnerability of poorer sections of society in general and farmers in particular. It is imperative to devise adaptive strategies to mitigate these vulnerabilities by ensuring sustainable water resource management. Farmer friendly resource management regimes require adequate institutional framework in place. Ostrom (1990) proposed eight design principles for institutions governing commons which later got interpreted for irrigation systems (Ostorm, 1992). Keeping these principles at the backdrop, the present paper uses case study method to document a community managed canal irrigation system, known as Allahrakhi, in Markham Grant of Dehradun District in Uttarakhand, India. It is important to analyse current community managed resource systems keeping researched

* This paper draws on field visits with Shalini Agarwal.

design principles in view so that the existing systems can be improved. This kind of analysis is also likely to inform people/ organizations that are helping or are likely to help communities to evolve their own indigenous systems.

Introduction

Water, as most of the other natural resources is available as Common Pool Resources. Commons have always been besotted with a variety of problems, but the emerging problems such as global climate change, as Stavins (2010) puts, are 'both more important and more difficult than problems of the past"(p.20). Natural resource systems, which are inherently characterized by a number of uncertainties, are being exacerbated by climate change (Nichols et al, 2011). Using observational records and climate projections, IPCC Technical Paper VI focusing on Climate change and Water reiterates the increasing vulnerability of freshwater resources due to climate change (Bates, Kundzewicz, and Palutikof, 2008). The authors further note that "Changes in water quantity and quality due to climate change are expected to affect food availability, stability, access and utilization. This is expected to lead to decreased food security and increased vulnerability of poor rural farmers...(p.3)". This observation makes it imperative to devise adaptive strategies to mitigate vulnerabilities of the poor farmers who depend on community waters for their survival. While efforts have been made to explore linkages between natural resource management and climate adaptive strategies at different levels (*e.g.*, Corfee-Morlot,., Berg M., and Caspar, G., 2003), sustainable natural resource management is not just about suitable technology and pricing. It is also about, as Ali (2010) observes, the institutions involved in resource management at the local level. He further notes that "policies of transferring management responsibility from the state to users have become increasingly widespread in response to the performance deficiencies of government agencies, the economic crisis of the state, and broader policies of decentralization (p. 2)."

Farmer friendly resource management regimes require adequate institutional framework for sustainable natural resource management. Based on her extensive field surveys, Ostrom (1990) proposed eight design principles for institutions governing commons which later got interpreted for irrigation systems (Ostorm,

1992). These principles included: Well-defined boundaries, Congruence between appropriation and provision rules and local conditions, Collective-choice arrangements, Monitoring, Graduated sanctions, Conflict-resolution mechanisms, Minimum recognition of rights, and Nested enterprises. Cox, Arnold and Tomas (2010) evaluated 91 studies which have explicitly or implicitly evaluated these 8 design principles. They found that the principles are well supported empirically and provided a reformulation of the design principles which are being summarized in the Annexure 1. It is important to analyse current community managed resource systems keeping these principles in view so that the existing systems can be improved. This kind of analysis is also likely to inform people/organizations who are helping or are likely to help communities to evolve their own indigenous systems. This paper presents a case study of a community managed canal irrigation system in the light of design principles outlined by Ostorm (1990,1992) and reformulated by Cox *et al.* (2010).

Methodology

Study Area

Lying at the foothills of the middle Himalayas, in between the streams of Song and Suswa, Markham Grant is situated in Doiwala block, Dehradun, Uttarakhand, India. Markham Grant is predominantly an agricultural community. Demographically, it is inhabited by Hindus towards the north, Sikhs in Centre and Muslims towards South in the irrigated area of Allahrakhi. There is absence of local Garhwali population in the area mainly because of the fact that the grant of Markham was made by the British in 1838 to ex-army men or the outsider for their contribution to the British Raj. The Allahrakhi had been constructed by them after the grant of the land and had been irrigating the lands since then.

Method

We have used an explanatory case study research method to explain the CPR management by studying the processes critical to water resource management. According to Yin, (1989), case refers to an event, an entity, an individual or even a unit of analysis. We have considered complete Allahrakhi canal irrigation system as one unit. The data collection included in-depth interviews with the various stakeholders over a period of two months and also participant observation of various processes involved in water management.

Results and Discussions

The Case

The sowing season for paddy had just started and as Jal Prabhandak (water manager) walked besides the Allahrakhi canal he felt the anxiety of managing the community owned canal irrigation system. He thought of the increasing pressures from the fellow irrigators, populist policies of the government and the unreliable monsoon. It was only in the last week during the annual maintenance of the canal that they had to call an emergency meeting of the Jal Samiti (water committee). One of the household had sent a young girl and not the ablest male member, as is the practice of the community, for the maintenance work. This was resented by other households as a fit case of free-riding. nThe girl's father quoted the example of a small land holder who did not even contribute even a single day of labour and yet no action was taken against them. The practice in the Chandmari branch of the irrigation system, a household owning less than two bighas (1 hectares = 6.25 Bighas) of land has to contribute one labour-day and if the holding is more than five bighas, the households have to contribute labour for the entire duration of the maintenance work. All household users have to share nominal cost of maintenance in proportion to their land holdings. After this dispute, the Jal Pradhan seriously thought of codifying the interpretation so as to prevent a rather subjective interpretation of the rules.

Located in the vicinity of Dehradun city, the capital of Uttarakhand state, Markham Grant has not been left untouched by the increasing commercialization of the erstwhile subsistence economy. Being close to the national highway has also brought increasing mechanization and commercialization of the agriculture in Markham Grant. With mechanization of agriculture and lucrative prices of sugarcane (in fact subsidized by the government) his fellow villagers have become greedy and area under sugarcane cultivation has drastically increased.

With increasing pressure on land due to increasing population, the water resources are also strained. The disputes and altercations on the issue of priority as well as quantity of irrigation water are increasing. Their emergency conflict resolution meetings are becoming less successful in arriving at results. Although users have threatened to go to the civil courts for resolution of the water sharing disputes but he has been spared the embarrassment till now. This could be troublesome as the management committee of the Allahrakhi canal is not recognized by the government.

The Jal Pradhan at times is financially helpless as all grants/aid/schemes of the government for improvement of irrigation facilities for his area are routed through the Panchayat Samiti or in fact through the Gram Pradhan (village head) himself. This creates resentment among different religious and caste groups as the subsequent allocation of the grants takes political and communal colours. Funds are also available to strengthen the agricultural infrastructure by making the canals in the area, including Allahrakhi, fully lined. The non-lined nature of the Allahrakhi Canal leads to loss of efficiency in carrying capacity of the water as the efficiency of unlined canals is only 60 per cent as compared to 70-75 per cent of fully lined canals.

Jal Pradhan understands fully well that an efficient management of common property resources (CPRs) like water requires clearly defined boundaries. Therefore, the rights to use the water are linked to land holdings and the water supplied through the canal is directly proportional to the land holdings. For example, Chandmari village which is at the head of a sub branch of the canal follows a simple water allocation rule- one hour of sub canal water for ten bighas of land. The organization of the whole system has been kept simple. A work team of four or five households form the first tier. The second tier may be of all the users of a particular branch of an irrigation system. The third tier is the collection of all the households of one headwork. The small teams prevent free riding and large enterprises provide economies of scale and aggregation of capital.

The Jal Prabhandak has carried forward the intergenerational collective choice arrangements for the use of irrigation water. In fact, the very spirit of community managed CPRs asks for inclusion of all the stakeholders in the decision making body. The absence of a statuary authority makes it imperative for the users to formulate effective rules which are in consonance with specific characteristics of their settings. In the Allahrakhi canal management, all decisions are taken by consensus and all monitoring is done by the users themselves. They had explored the idea of an external/non-user agent for monitoring purposes but rejected the idea by consensus. They felt that such an official would require some compensation to catch and/or report a malpractice by any of the users. Also Jalprabhandak takes care of monitoring the rotation of irrigation water among the users of particular branch. For the management of actual usage, it is almost self regulatory and violations of the operational rules in community managed canal irrigation system at Markham Grant are met with graduated sanctions. In the case of Markham grant, for example, in the water rotation system, the irrigator who nears the end of a rotation would like to extend the time of his turn (and thus extend the amount of water obtained) but the next irrigator in the rotation system waits for him to finish and would like to start early. Thus the presence of first irrigator deters the second from starting early and that of second form ending late for the first.

Increasing government involvement has brought unheard of situations which their forefathers had neither faced nor had thought of them. Recently landless villagers have started asking for a share in the water rights of the canals as is the case with some of the crafted institutions in the country which is opposed by traditional right holders. The land holders view this as a ploy to create dissent in the community. They feel that these landless villagers already use the canal water for non- irrigation purposes like cattle rearing, washing, cleaning etc without contributing to the maintenance etc. Some of the smaller landholders and landless villagers with covert support of Gram Pradhan (village head) have started clamoring for taking the support from the government.

More trouble is brewing in the neighbourhood for the community and the canal. The pressure created by the civil rights group has forced the government to develop another site for rehabilitation of the dam- displaced people. The new site would be located close to the Markham Grant as the land is fertile and the perennial source of water is available in the form of Song River. There is talk of getting another government canal from the river and it could be sourced closer to the origin of the rivulet from where Allahrakhi also draws water.

He remembered meeting one of the government dignitaries, few year back, who wanted to replicate this model across the length and breadth of the country. He had to tell him bluntly that we are a small and cohesive community; it may not be possible to replicate the same process in every village of the country. In the changing scenario, he is not sure how long would he be able to call themselves as a 'small and cohesive community' when the pressures of commerce and politics are tearing at the fabric of their community. The monsoon has been pretty predictable and generous this year but Jal Pradhan wondered how long can they manage the growing needs and aspirations of the users of the Allahrakhi? Would he be forced to see the privatization or takeover of the Allahrakhi by the government?

Some Observations

It is apparent that the community based management system is working reasonably well. The principles summarized in Annexure-1 seem to be working in Allahrakhi. However, the challenges identified in literature also seem to be at work in the case under discussion. For example, Klooster (2000), as cited by Cox et al (2010) asserts that development and market forces can destabilize CPR arrangements that worked well when the community and resource were isolate. It is evident that mechanization, changes in economy, growing population pressure and growing connectivity may not keep the community 'small and cohesive' as Jal Prabandhak laments in the case.

In consonance with Principle 7, external government agencies do not challenge the right of local users to create their own institutions which, in some instances, goes against the well being of the community managed system. For example, as noted in the case, the Jal Samiti is not entitled to avail benefits of any government scheme meant to improve irrigation systems.

The appropriation provision rules need to be codified for improving the compliance. At the same time, conflict resolution mechanism particularly between the users' representatives and other agencies such as Village Panchayat, other civil society organizations and government officials, need to be put in place.

Conclusion

It is important to note, as Laerhoven, and Ostrom (2007) puts it, that 'there is a world of nuances between the State and the

Market.(p.19)". Principles are not 'magic bullet or institutional panacea' and as Cox et al (2010) suggested 'a nested typologies of environmental problems and governance arrangements' are needed. Any one or a combination of privatisation, nationalisation, and community management could be employed and any of these three alternative systems could be as good or as bad as any other. (Singh, 2009). Despite Hardin's dominant discourse of ' tragedy of commons', it has been repeatedly found that under certain conditions, communities given the right to self-organise can govern themselves democratically and environmentally sustainable way. However, one needs to keep in mind the cautionary note put forth by Harkes (2006) that Ostorm's principles are characteristics of the community or institution, but" the real 'glue' that keeps an institution alive over time are the social mechanisms, i.e., trust, legitimacy, and transparency (p.251)." Narain (2009) also argues for a property rights framework, which encourages cooperation. There have been concerns over achieving climate and water justice which requires efficient, effective and just institutional mechanisms. Against this backdrop, it is also worthwhile to test all some of the principles for other institutional arrangements.

References

Ali, A. (2010). *Natural Resources Management and Climate Change: Policy and Institutional Arrangement*, A Paper Compiled and Prepared for Africa Local Government

Action Forum (ALGAF X Session II).Addis Ababa: Municipal Development Partnership/The World Bank institute

Bates, B.C., Z.W. Kundzewicz, S. Wu and J.P. Palutikof, Eds., (2008). *Climate Change and Water.* Technical Paper of the Intergovernmental Panel on Climate Change, IPCC Secretariat, Geneva, 210 pp.

Corfee-Morlot,J., Berg M., and Caspar, G.(2003).*Exploring linkages between natural resource management and climate adaptation strategies*, Paris: OECD.

Cox, M., Arnold, G., and Tomás, S.V., (2010). A review of design principles for community-based natural resource management. *Ecology and Society 15(4): 38.*

Hardin, G. 1968. The Tragedy of the Commons. *Science*, 162:1243-1248.

Harkes, I. H. T. 2006. *Fisheries co-management, the role of local institutions and decentralisation in Southeast Asia: with specific reference to marine sasi in Central Maluku, Indonesia.* Dissertation. Leiden University, Leiden, The Netherlands.

Laerhoven, F., and Ostrom, E. (2007). Traditions and Trends in the Study of the Commons. International Journal of the Commons. 1(1). 3-28

Klooster, D (2000). Institutional choice, community, and struggle: a case study of forest co-management in Mexico. *World Development* 28(1):1-20.

Narain, S. (2009). A Just Climate Agreement: The Framework for an Effective Global Deal. In S. Narain, P. Ghosh, N C Saxena, J. Parekh and P. Soni. *Climate Change - Perspectives from India. Delhi: UNDP India.*

Ostrom, E. 1990. *Governing the Commons: The Evolution of Institutions for Collective Action.* Cambridge University Press, Cambridge, UK.

Ostrom, E. (1992). Crafting *Institutions for Self-Governing Irrigation Systems.* San Francisco: Institute for Contemporary Studies Press.

Nichols, J.D., Koneff, M. D. Helgund, P.J., Knutson, M.G., Seamans, M.E,.,Lyons, J.E., Morton, J.M. Jones, M.T., Boomer, G.S., and Williams, B.K., (2011). Climate Change, Uncertainty, and Natural Resource Management. *Journal of Wildlife Management* 75(1):6–18;

Singh, K (Feb. 2009). *Improving Efficiency and Sustainability of India's Agriculture through Judicious Management of Common Property Resources : A Perspective.* Paper presented in Fourth World Congress on Conservation Agriculture. Delhi.

Stavins, R.N. (2010). *The Problem of the Commons: Still Unsettled After 100 Years. Discussion Paper.* Washington: Resources for the future.

Yin, R., (1989). *Case Study Research.* Sage Publication, California, pp: 22-26.

Appendix 1
List of Proposed Modified Design Principles
(Source: Cox, Arnold, and Tomás, 2010)

Principle	*Description*
1A	User boundaries: Clear boundaries between legitimate users and nonusers must be clearly defined.
1B	Resource boundaries: Clear boundaries are present that define a resource system and separate it from the larger biophysical environment
2A	Congruence with local conditions: Appropriation and provision rules are congruent with local social and environmental conditions
2B	Appropriation and provision: The benefits obtained by users from a common-pool resource (CPR), as determined by appropriation rules, are proportional to the amount of inputs required in the form of labor, material, or money, as determined by provision rules.
3	Collective-choice arrangements: Most individuals affected by the operational rules can participate in modifying the operational rules.
4A	Monitoring users: Monitors who are accountable to the users monitor the appropriation and provision levels of the users.
4B	Monitoring the resource: Monitors who are accountable to the users monitor the condition of the resource.
5	Graduated sanctions: Appropriators who violate operational rules are likely to be assessed graduated sanctions (depending on the seriousness and the context of the offense) by other appropriators, by officials accountable to the appropriators, or by both.
6	Conflict-resolution mechanisms: Appropriators and their officials have rapid access to low-cost local arenas to resolve conflicts among appropriators or between appropriators and officials.
7	Minimal recognition of rights to organize: The rights of appropriators to devise their own institutions are not challenged by external governmental authorities.
8	Nested enterprises: Appropriation, provision, monitoring, enforcement, conflict resolution, and governance activities are organized in multiple layers of nested enterprises.

Chapter 20

Generation of Hydro-electricity: A Case Study of Himachal Pradesh

☆ *Sujit Surroch and Anita Surroch*

Mankind has identified couples of renewable and non-renewable sources of energy. Hydro (water) is one of the main and globally harnessed renewable sources. It is a flexible source of electricity since plants can be ramped up and down very quickly to adapt to changing energy demands. However, damming interrupts the flow of rivers and can harm local ecosystems, and building large dams and reservoirs often involves displacing people and wildlife and requires significant amounts of carbon-intensive cement. Once a hydro-electric complex is constructed, the project produces no direct waste, and has a considerably lower output level of the gas carbon dioxide, fossil fuel and generated through the use of the gravitational force of falling or flowing water is known as hydro-electricity.

Hydro-electricity: A Global Phenomenon

Hydro-electricity is the most widely used form of renewable energy, accounting for 16 percent of global electricity consumption, and 3,427 terawatt-hours of electricity production as in 2010. It is produced in 150 countries, with the Asia-Pacific region generating 32 percent of global hydropower. China is the largest hydro-electricity producer, with 721 terawatt-hours of production in 2010, representing around 17 percent of domestic electricity use. Each of three Gorges Dam in China, Itaipu Dam in Brazil, and Guri Dam in Venezuela generate more than 10 GWs hydro-electricity. The cost of hydro-electricity is relatively low, making it a competitive source of renewable electricity. India has a potential of 150,000 MW of which less than one fifth is being tapped.

Process of Generating Hydro-electricity

Most hydro-electric power comes from the potential energy of dammed water driving a water turbine and generator. The power extracted from the water depends on twin factors- the volume of water and height between the source and the water's outflow. The difference between source and outflow is called the head. The amount of potential energy in water is proportional to the head. A large pipe (the "penstock") delivers water to the turbine. Power generated from water is termed as Hydro-electricity. Hydro-electricity means electricity generated by hydropower or by using gravitational force of falling or flowing water. This form of energy neither produces any

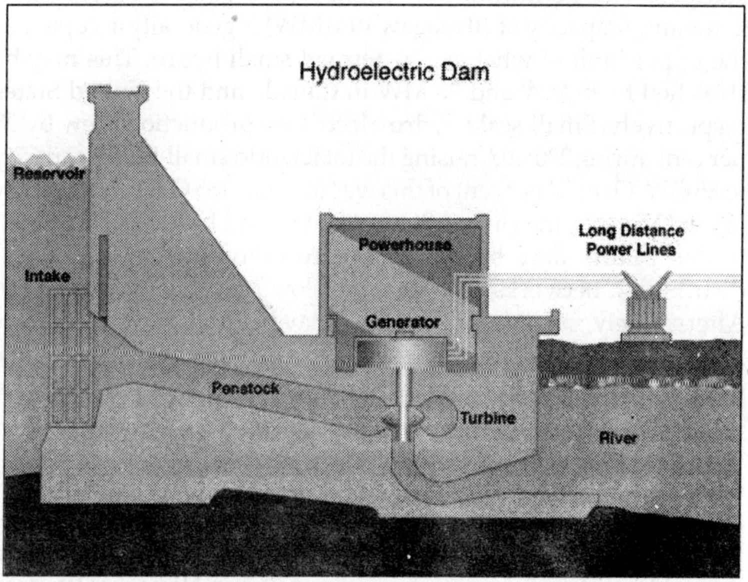

direct waste matter nor is subjected to exhaustion. In order to have better understanding and deeper insight of the way hydro-electricity is generated a picture of hydro-electric dam is being given.

Classification of Hydro-electric Facilities

There is no uniformly accepted range of sizes of hydropower stations. However for academic purposes and based on current trend of classification brief description of large, small, micro and pico hydropower capacities is being as under;

Large

Although no official definition exists for the capacity range of large hydro-electric power stations, facilities from over a few hundred megawatts to more than 10 GW are generally considered large hydro-electric facilities. In other words power stations which are not mini, micro and small and are larger than all of these are termed as large.

Small

In India, hydro power projects with a capacity of up to 25 megawatt (MW) fall under the category of small hydro power (SHP) stations. The definition of a small hydro project varies but a generating capacity of 10 megawatts (MW) is generally accepted as the upper limit of what can be termed small hydro. This may be stretched to 25 MW and 30 MW in Canada and the United States respectively. Small-scale hydro-electric ity production grew by 28 per cent during 2005-08, raising the total world small-hydro capacity to 85 GW. Over 70 per cent of this was in China (65 GW), followed by Japan (3.5 GW), the United States (3 GW), and India (2 GW). Small hydro plants may be connected to conventional electrical distribution networks as a source of low-cost renewable energy. Alternatively, small hydro projects may be built in isolated areas that would be uneconomic to serve from a network, or in areas where there is no national electrical distribution network. Since small hydro projects usually have minimal reservoirs and civil construction work, they are seen as having a relatively low environmental impact compared to large hydro. This decreased environmental impact depends strongly on the balance between stream flow and power production. As many as 294 mini and micro hydropower projects in the public and private sectors are coming up in Himachal Pradesh to generate 586 MW of electricity. Mini and micro hydro-electric power projects were allotted to private and joint venture companies and Himurja (state energy development agency). The tentative cost of all these projects is Rs.4,104.31 crores and these projects would generate 586 MW of power. Most of the projects are located in Kullu,

Shimla, Lahaul and Spiti, Mandi, Sirmaur and Kangra districts. Kullu alone has 83 projects.

Micro

It is a term used for hydro-electric power installations that typically produce up to 100 KW of power. These installations can provide power to an isolated home or small community, or are sometimes connected to electric power networks. There are many of these installations around the world, particularly in developing nations as they can provide an economical source of energy without purchase of fuel. Micro hydro systems complement photovoltaic solar energy systems because in many areas, water flow, and thus available hydro power, is highest in the winter when solar energy is at a minimum.

Pico

Under 5 KW of hydro-electric power is generated in it. It is useful in small, remote communities that require only a small amount of electricity. For example, it may cater to the needs to power one or two fluorescent light bulbs and a TV or radio for a few homes. Even smaller turbines of 200-300W may power a single home in a hill state like Himachal Pradesh with a drop of only 1 m (3 ft). Pico-hydro setups typically are run-of- the-river which means that dams are not used, but pipes divert some of the flow, drop this down a gradient, and through the turbine before returning it to the stream.

State of Hydro-electricity in Himachal Pradesh

State of Himachal Pradesh is nestled in the western Himalayas and has great potential for hydro-electricity generation. It is one of the states that have most actively implemented the Hilly Hydro Programme. Himachal Pradesh is endowed with five river basins, namely Yamuna, Satluj, Beas, Ravi and Chenab. All the five rivers originate from the Himalayas and are perennial and cover the entire state running from East to West and North to South. Through preliminary investigations, it has been estimated that about 23,194.95 MW of hydel potential can be exploited in the state by constructing various major, medium, small and mini/micro hydel projects on these five river basins. Out of this hydel potential 6728 MW has been harnessed by central and state government agencies and private players till 2011. The state government has framed a four pronged strategy under which projects are to be executed under various management modules *i.e.* in the State Sector, Joint Sector, Private Sector and the Central Sector to exploit this huge potential and to give utmost thrust to the development of hydro-power.

Priority is being accorded for speedy harnessing of balance hydro power potential in a manner that optimizes returns to the State. With a view to achieve this, the State initiated the process of involving both the Central Government PSUs and the Private Sector in a big way in hydro-power development. The proposed distribution of above 23000 MW potential in the state is likely to be 17 per cent by State sector, 39 per cent by Central/Joint sector, 38 per cent by private sector and remaining 6 per cent comprise of projects with a capacity of less than 5 MW.

Potential and Generation of Hydropower in HP

The state government has been according the highest priority for its development so that hydel generation meets the growing need of power for industry agriculture and rural electrification. A methodology was developed to identify potential sites using available hydrological data, topographical maps and satellite information. Thus as many as 2000 potential sites have been identified in the state. A soft ware called HydrA-HP has been prepared by the alternate Hydro Energy Centre in Roorkee (India) and the Centre for Ecology and Hydrology in Wallingford (UK). This has prepared a model of river flow in the water courses in the state and provides a rapid means of estimating the hydropower potential of any location in the state. It reduces the initial expenditure on hydrological surveys to be conducted by the developers of projects.

Basin-wise Potential

It is discernible from the information compressed in Table 20.1 that of the total assessed potential of hydro power in the state nearly half (48 per cent) have been identified in the Satluj river basin, more than 22 per cent In Beas and a little more than 11 per cent in Ravi and Chenab each. More than 85 per cent of the electricity is being generated by joint sector. Share of state and private sectors is about 7 per cent and 6 per cent respectively. More than half of the total electricity is being produced on Satluj river alone.

Table 20.1: Potential Assessed and Tapped

Sl.No.	Basin	Total Assessed Potential (MW)	Potential Installed (MW)		Per cent age of Total Installed to Total Assessed			
			State Sector	Private Sector	Central/ Joint Sector	Himurja	Total	
1.	Satluj	9866.55 (48.32)	50.25	300.00	2825.00	3.90	3279.15 (50.84)	33.23
2.	Beas	4226.90 (22.66)	226.50	02.00	496.00	56.95	881.45 (29.17)	40.66
3.	Ravi	2345.25 (0.48)	0.25	0.00	038.00	6.20	064.45 (6.50)	45.37
4.	Chenab	2251.00 (0.02)	0.00	0.00	0.00	0.00	0.00	0.00
5.	Yamuna	602.52 (2.95)	79.95	0.00	31.57	2.00	223.52 (3.46)	37.14
6.	Other small rivers/ Rivulets	723.40 (3.54)	0.00	0.00	0.00	0.00	0.00	0.00
	Total	20415.62	466.95 (7.24)	402.00 (6.23)	5490.57 (85.14)	89.05 (0.38)	6448.57	31.58

Note: Figures in parentheses show percentage in total.

Hydropower Generating Agencies

Several agencies have been embarking on the generation of hydropower in the state. Bhakhra Bias Management Board (BBMB-2711 MW), Satluj Jal Vidyut Nigam Limited (SJVNL-1500) and National Hydro Power Corporation (NHPC-1038 MW) respectively are the largest power producing organizations. Besides, NHPC Limited is surging ahead in developing hydro power in Himachal Pradesh, thereby changing the socio-economic conditions and augmenting overall development of the region. NHPC Limited has already commissioned three projects, *viz.* Baira Siul (180 MW), Chamera Stage – I (540 MW) and Chamera Stage – II (300 MW). Three more projects, *viz.* Parbati Stage – II (800 MW), Parbati Stage – III (520 MW) and Chamera Stage – III (231 MW) are under active construction stage.

Table 20.2: Executing Agency and Installed Capacity

Sl.No.	*Executing Agency*	*Installed Capacity*	*Percentage Share in Total*
1.	BBMB	2711.00	42.03
2.	PSEB	110.00	1.70
3.	NHPC	1038.00	16.09
4.	UPSEB	131.57	2.63
5.	SJVNL	1500.00	23.25
6.	HPSEB	466.95	7.22
7.	IPPs	402.00	6.23
8.	HIMURJA	89.05	1.38
	Total	6448.57	

The Himachal Pradesh Power Department, with the help of the Himachal Pradesh State Electricity Board Corporation, has done a commendable job of identifying 120 power projects with a capacity 17,000 MW (50 per cent projects would be of 5 to 50 MW) across five river basins of the state. Ninety-five per cent projects are run-off-the-river (RoR) type and are located across the rivers Sutlej, Beas, Ravi, Chenab, Pabbar (a tributary of Yamuna) and their tributaries. Out of 120 projects, at least 70 per cent have been allotted and these are under different stages of execution. Remaining 30 per cent projects (with a combined capacity of about 5000 MW) will be allotted soon.

These projects have been initiated both by private and public sector players (SVJN, NTPC, NHPC, HPPCL, HPSEB) almost in equal proportion. As such, by 2020, Himachal Pradesh will have an additional power generation of 17,000 MW (7650 MW for 24x7 power generation) and earn good revenue.

Contribution of Joint Sector

It is the major electricity generating agency in the state. With the production of 5491 MW out of the total of 6449 MW, it has been contributing about 85 per cent of the total power being produced.

Table 20.3: Projects and Installed Capacity

Sl.No.	*Name of the Project*	*Executing Agency*	*Installed Capacity (MW)*
1.	Yamuna Projects (HP Share)	Uttrakhand	131.57
2.	Baira Suil	NHPC	198.00
3.	Chamera-I	NHPC	540.00
4.	Chamera-II	NHPC	300.00
5.	Shanan Project	PSEB	110.00
6.	Pong Dam	BBMB	396.00
7.	Dehar	BBMB	990.00
8.	Bhakra	BBMB	1325.00
9.	Nathpa JHakri	SJVNL	1500.00
	Total		5490.57

Hydropower to be Generated by Himachalis

Small Hydro Project up to 2 MW capacity shall be exclusively be reserved for the Himachalis. Whereas while allotting projects up to 5 MW, Himachali shall be given extra 30 marks in addition to the marks obtained by them for financial strength. Most of the projects are located in the basins of the Sutlej and Chenab rivers in Lahaul and Spiti districts. The projects have been allotted on a build-own-operate-and-transfer (BOOT) basis. The fixed quota is 12 percent free power for the first 12 years, 18 percent for up to 30 years, and 30 percent up to 40 years. After 40 years, the project will be handed over to the government. The free power bidding is in addition to the fixed up front premium. The free power bidding is in addition to the fixed up front premium of Rs. 2 million (Rs.20 lakh) per megawatt. The government has also decided not to charge any up front premium from the state-based independent power producers (IPP) up to 5 MW in view of the development of entrepreneurship amongst the educated youth of Himachal Pradesh in the power sector. However, they will be required to provide free power to the state as per its policy.

Hundred Years of Hydro-electricity Generation

The Hydro-electricity development in Himachal Pradesh dates back to the pre-independence period. A humble beginning was made in 1908 when king of erstwhile Chamba princely state developed a small project of 0.45 MW capacity. The next hydel project at Chaba with capacity of 1.75 MW near Shimla was executed in 1912. Diverting water of Uhl rivulet by a tunnel at Barot yet another hydro-power project of 48 MW capacity was constructed at Shanan near Jogindernagar in District Mandi during first quarter of twentieth century.

Hydro Policy of state

In order to embark upon our ambitious programme in Hydel generation, the Pradesh had formulated its own Hydro-Power Policy in 2006 and has further improved this policy in the last three years. It seeks to safeguard the interest of the people of the State on one hand and to protect the delicate ecology and environment on the other. Mandatory provision of 1.5 per cent of project cost has been kept in the state policy for the development of local area during the construction stage. Besides, a provision for 1 per cent additional free power has been earmarked for development of local area in line with the National Hydro-Policy in post- commissioning scenario.

Hydro Policy on Privatization

The state government has formulated a Hydro Policy to attract private entrepreneurs to harness immense hydro potential in the state. Some of the main features of this policy are as under;

1. Selection of Developer on MoU Route allowed for projects up to 100 MW
2. Selection of Developer on ICB route for Projects above 100 MW.

3. No Clearances necessary from CEA for projects selected on competitive bidding route for projects costing up to Rs. 2500 crores.
4. Secondary energy rate to be at-par with primary energy. Premium on peak power proposed.
5. Process of Transferring clearances to IPP's simplified.
6. 100 per cent Foreign Equity permitted on the automatic approval route provided it does exceed Rs. 1500 Crores.
7. Limit of 40 per cent financing from Indian Financial Institutions waived off.
8. Tariff determination by SERC/CERC.
9. Projects upto 25 MW to be transferred to MNES.
10. HPSEB to purchase power @ Rs. 2.50/kWh from projects upto 5 MW capacity.
11. Banking and Wheeling Facilities permitted @ 2 per cent including transmission losses.
12. Incentives for small HEP as per GOI Guidelines.
13. Moratorium of 5 years for payment of Electricity duty by companies which consume electricity produced by them.

Thus the state government has taken an initiative to give boost to generate hydro-electricity on the rivers, rivulets and other water bodies.

Private Sector Participation

The State Government has taken several initiatives to encourage private sector participation in small hydro power development. Himachal Pradesh is among the few States, which has streamlined and is continuously refining various procedures/processes to minimize the bottlenecks. The process of exploitation of hydel potential in small hydro sector through private sector participation began during 1995-96. Since then, the allotment of project sites has been a continuous process. Till 31st March 2010, 509 Small hydro Electric Projects (up to 5 MW capacity) with an aggregate capacity of 1,299 MW have been allotted. Out of these, 29 projects with an aggregate capacity of 114.50 MW have already been commissioned.

Table 20.4: Projects Under Private Sector Execution

Project	*Capacity*	*M/S*	*Estimated Cost*	*Date of Signing the MoU*	*Date of Commossioning*
Neogal Hydel Project	5 MW	Om Power Corporation Ltd., New Delhi	Rs. 61.74 crore	28.8.93	2010-
Allain Duhangan Hydel Project	92 MW	Rajasthan Spinning and Weaving Mills Ltd., New Delhi	Rs. 922.36 crore	22.2.200	2010-11
Karcham-Wangtoo Hydel Project	000 MW	Karcham Hydro Corporation Ltd., New Delhi and Jai Parkash Industries Ltd.	Rs. 15910 crore	28.8.93	2011-12
Malana-II	00 MW	Everest Power Pvt. Ltd., New Delhi	Rs. 633.47 crore	27.5.2002 and 4.1.2003	2011-12
Tangnu Romai Stage-I and II	44MW + 6 MW	P.C.P International Ltd.	Rs. 239.73 crore	5.7.2002	2012-13
Lambadug	25 MW	Himachal Consortium Power Projects Pvt. Ltd.,	Rs. 49.81 crore	4.6.2002	2012-13
Budhil Hydro Electric Project	70MW	Lanco Green Power Pvt. Ltd.	Rs. 418.80 crore	23.9.04	2011-12
Sorang Hydro Electric Project	00MW	Himachal Sorang Power Pvt. Ltd.	Rs. 1586 crores	23.9.2004 and 28.1.2006	2011-12
Tidong-I Hydro Electric Project	00MW	Nuziveedu Seeds Ltd., Secunderabad	Rs. 500.11 crores	23.9.2004 and 28.7.2006	2013-14
Fozal Hydro Electric Project	9 MW	Fozal Power Pvt. Ltd., New Delhi	Rs. 49.17 crores	21.6.2000 and 3.4.2006	2012-13

Contd...

Table 20.4–Contd...

Project	*Capacity*	*M/S*	*Estimated Cost*	*Date of Signing the MoU*	*Date of Commossioning*
Paudital Lassa Hydro Electric Project	24 MW	Shree Lay Lakshami Power Co. Ltd.	Rs. 107.40 crore	6.6.2002 and 26.10.2006	2011-12
Baragaon Hydro Electric Project	24 MW	Kanchanjunga Power Pvt. Ltd., Noida (U.P.)	Rs. 168.09 crores	6.6.2002 and 25.11.2006	2012-13
Baner-II Hydro Electric Project	6 MW	Prodigy Hydro Power (P) Ltd.	Rs. 30.36 crores	29.5.2000 and .10.200	2012-13
Raura Hydro Electric Project	8 MW	DLI Power (India) Pvt. Ltd., Pune	Rs. 42.03 crores	4.2.1996 and 24.3.2008	2012-13
Sai Kothi Hydro Electric Project	5 MW	Venture Energy and Technologies Ltd., New Delhi	Rs. 83.17 crores	June, 2002 and 6.7.2007	2012-13
Patikari	6MW	Patikari Power Private Ltd., New Delhi	Rs. 125.9 crore	21.6.2000 and 9.11.200	2008
Kut	24 MW	Kut Energy (P) Ltd., Noida, U.P.	Rs. 196.5 crore	28.4.2007 and 25.5.2008	2011-12
Harsar	60MW	Soffimat SA France, Sobra Power Pvt. Ltd., Chandigarh	–	21.6.2000	2015-16
Bharmour	45MW	Soffimat SA France, Sobra Power Pvt. Ltd., Chandigarh	–	21.6.2000	2012-16

International Funding to State

Being live to the fact that Himachal Pradesh, a small mountainous state with five major rivers, has about a quarter of India's total potential hydropower resources following international agencies have provided loan to the state.

1. *World Bank* approved Climate Investment Funds for India to the tune of US$775 million. It includes the amount earmarked as Himachal Pradesh Development Policy Loan to the tune of US$200 million.
2. *The Asian Development Bank* (ADB) is providing loans of up to $350 million for upgrading of transmission system of Himachal Pradesh. The ADB Board of Directors has approved the financing facility for the Himachal Pradesh Clean Energy Transmission Investment Program, with a first installment of $113 million. It will help expand the supply of power, both in and outside the state, as well as promoting job opportunities, increased social services and poverty reduction.

Hydropower Mission 2020

The main objective of the Development Policy Loan is be to accelerate hydro power development in the state. As mentioned in the India's Climate Investment Plan; "Himachal Pradesh is one of the leading hydropower states in India and hydropower (large and small) by far dominates its renewable energy resource potential. Himachal Pradesh is among the few states which has streamlined and crystallized various procedures to minimize the bottlenecks and has come up with an investor friendly hydropower policy to attract private sector investment. The intent would be to implement a framework that would spur hydropower development to the tune of 70 percent by 2020. It would be a remarkable achievement that could serve as a model for other hydropower states in the country, and in the South Asia Region. The project presents a strong CO_2 emissions reduction potential." It is very clear that the World Bank is giving a Rs. 1000 crore blank check to the Himachal Pradesh Government to accelerate hydro development.

Advantages of Hydro-electricity

Low Power Costs

Hydropower is cheap, non-polluting, and many small facilities (less than 30 MW) are considered renewable energy sources. Unlike other sources of energy like nuclear power, the costs of large scale plants is declining because the facilities have been paid for by federal or state projects. Dams and reservoirs also provide recreational opportunities, flood control, irrigation, refuges for wildlife, water supply, and transportation. The cost of operating a hydro-electric plant is nearly immune to increases in the cost of fossil fuels such as oil, natural gas or coal, and no imports are needed. The average cost of electricity from a hydro plant larger than 10 megawatts is 3 to 5 U.S. cents per kilowatt-hour.

Long Economic Lives

Hydro-electric plants have long economic lives, with some plants still in service after 50–100 years.

Poverty Alleviation

Harnessing hydel energy can be an effective way of reducing poverty. China and India have already done that successfully. It is not a fruitful idea to keep discussing as to how much the number of people living below the poverty line has come down during the last five years. Considering that even with the reduced percentage, those suffering from acute poverty are still too many to be left on their own.

Reduction in Pollutants

Since hydro-electric dams do not burn fossil fuels, they do not directly produce carbon dioxide. While some carbon dioxide is produced during manufacture and construction of the project, this is a tiny fraction of the operating emissions of equivalent fossil-fuel electricity generation.

Other Uses of the Reservoir

Reservoirs created by hydro-electric schemes often provide facilities for water sports, and become tourist attractions themselves. In some countries, aquaculture in reservoirs is common. Multi-use dams installed for irrigation support agriculture with a relatively constant water supply. Large hydro dams can control floods, which would otherwise affect people living downstream of the project.

In nutshell, fuel should not be burned to minimize pollution. Water to run the power plant is being provided free by nature. Hydropower plays a major role in reducing greenhouse gas emissions. And relatively low operations and maintenance costs. The technology is reliable and proven over the time.

Disadvantages of Hydro-electricity

Hydro-electricity generation is pregnant with multiplicity of disadvantages. Some of the main disadvantages are as follows:

Siltation and Flow Shortage

When water flows it has the ability to transport particles heavier than itself downstream. This has a negative effect on dams and subsequently their power stations, particularly those on rivers or within catchment areas with high siltation. Siltation can fill a reservoir and reduce its capacity to control floods along with causing additional horizontal pressure on the upstream portion of the dam. Eventually, some reservoirs can become completely full of sediment and useless or over-top during a flood and fail.

Methane Emissions

Lower positive impacts are found in the tropical regions, as it has been noted that the reservoirs of power plants in tropical regions may produce substantial amounts of methane. This is due to plant material in flooded areas decaying in an anaerobic environment, and forming methane, a potent greenhouse gas. According to the World Commission on Dams report, where the reservoir is large compared to the generating capacity (less than 100 watts per square metre of surface area), there is possibility of omitting methane.

Rehabilitation

Another disadvantage of hydro-electric dams is slow rehabilitation relocate of people living where the reservoirs are planned. According to WCD estimated 40-80 million people worldwide had been physically displaced as a direct result of dam construction. Historically and culturally important sites can be flooded and lost. Himachal Pradesh is also not an exception. In H.P. first two major dams *i.e.* Pong relocated more than one lakh families in mid 70s and then Chamera-I, II and other dams relocated more than thousand families. They are still waiting to be rehabilitated forthe more than 5 decades.

Hydropower and the Environment

Hydropower does not pollute the water or the air. However, hydropower project can have large environmental impacts by changing the environment and affecting land use, homes, and natural habitats in the dam area. Most hydro-electric power plants have a dam and a reservoir. These structures may obstruct fish migration and affect their populations. Operating a hydro-electric power plant may also change the water temperature and the river's flow. These changes may harm native plants and animals in the river and on land. Reservoirs may cover people's homes, important natural areas, agricultural land, and archaeological sites. So building dams can require relocating people. Methane, a strong greenhouse gas, may also form in some reservoirs and be emitted to the atmosphere.

In fact hydro-electricity eliminates the flue gas emissions from fossil fuel combustion, including pollutants such as sulfur dioxide, nitric oxide, carbon monoxide, dust, and mercury in the coal. Hydro-electricity also avoids the hazards of coal mining and the indirect health effects of coal emissions. Compared to nuclear power, hydro-electricity generates no nuclear waste, has none of the dangers associated with uranium mining, nor nuclear leaks. Unlike uranium, hydro-electricity is also a renewable energy source. Compared to wind farms, hydro-electricity power plants have a more predictable load factor. If the project has a storage reservoir, it can generate power when needed. Hydro-electric plants can be easily regulated to follow variations in power demand. Although hydropower is non-polluting, construction of new facilities is highly controversial because of environmental and social impacts. The best location for plants is often in mountainous or rainy areas where ecosystems are fragile and often protected. New technology along with upgrades in current plants can increase efficiency and create energy sources without disrupting more lands.

In financial year 2011-2012, 2090 MW power with the commissioning of the 1000 MW Karcham Wangtoo project, Malana-II (100 MW), Budhil(70 MW), Chamera-III (231 MW), Parvati (520 MW) and some other small Hydel Projects was likely to be added in the power system. It has been targeted that by the end of the 12th Plan the State would have over 17,000 MW of generating capacity. The state has already achieved unique distinction of electrification of 100 per cent revenue villages of the state.

References

Worldwatch Institute (January 2012). "Use and Capacity of Global Hydropower Increases". http://www.worldwatch.org/node/9527.

Renewables 2011 Global Status Report, page 25, Hydropower, *REN21*, published 2011, accessed 2011-11-7.

"Pico Hydro Power". T4cd.org. http://www.t4cd.org/Resources/ICT_Resources/Projects/Pages/ICTProject_287.aspx. Retrieved 2011-09-16.

Hydropower – A Way of Becoming Independent of Fossil Energy?

"Beyond Three Gorges in China". Waterpowermagazine.com. 2011-08-10.

"Sedimentation Problems with Dams". Internationalrivers.org. http://internationalrivers.org/en/node/1476. Retrieved 2011-12-16.

Economic Survey Himachal Pradesh various issue

Draft Annual Plan 2010-11.

Balokhra, Jag Mohan (1997), The Wonderland Himachal Pradesh, H G Publications New Delhi.

www.worldbank.org.in/hydropower

www.worldbank.org.in

Frauke Urban and Tom Mitchell 2011. Climate change, disasters and electricity generation. London: Overseas Development Institute and Institute of Development Studies

"Briefing of World Commission on Dams". Internationalrivers.org. 2008-02-29. http://internationalrivers.org/en/way-forward/world-commission-dams/world-commission-dams-framework-brief-introduction.

Batta, R.N. (2000). *Tourism and the Environment: A Quest for Sustainability*. Indus Publishing. http://books.google.com/?id=lDJrdjwM7qYC and dq=himachal. Retrieved 2012-03-11.

Singh, Mian Goverdhan (2000). C.L. Gupta and Kulbhushan Chandel. ed. *Himachal Pradesh: History, Culture and Economy* (extensively revised ed.). Shimla.: Minerva Publishers and Distributors.

Singh, Sarina (2005). *India*. Lonely Planet. http://books.google.com/?id=Fk8FQa2ZSFQC and dq=himachal+history. Retrieved 2008-03-31.

Tiwari, A.K. (2000). *Infrastructure and Economic Development in Himachal Pradesh*. Indus Publishing. http://books.google.com/?id=aaXQwI5j9-MC and dq=himachal. Retrieved 2008-03-31.

Statistics and Data, Planning Department, Government of Himachal Pradesh

Chapter 21

Hydro-electric Power Generation in India: An Analysis

☆ *Sujit Surroch*

India has been endowed with bountiful natural resources. Hydropower, a renewable energy, is one of such resources. It is mechanism in which water cycle of Earth is used to generate electricity. Water evaporates from the Earth's surface, forms clouds, precipitates back to earth, and flows toward the ocean. The movement of water as it flows downstream creates kinetic energy that can be converted into electricity. Hydropower supplies at least 50 per cent of electricity production in 66 countries and at least 90 per cent in 24 countries. The word hydro means water. Hydropower is the energy that comes from force of moving water. Hydropower also known as hydrokinetic power or water power is power that is derived from the force or energy of falling water. Since ancient times, hydropower has been used for irrigation and the operation of various mechanical devices, such as watermills, sawmills, textile mills, dock cranes, and domestic lifts. Since the early 20th century, the term is used almost exclusively in conjunction with the modern development of hydro-electric power.

Hydropower Potential in India

India has been using hydropower for hundreds of years building water wheels and watermills. As per recent estimates, India is blessed with immense amount of hydro-electric potential and ranks 5th in terms of exploitable hydro-potential on global scenario. India has multiple river systems flowing through its length and breadth. These include six major river systems, namely, the Indus, Brahmaputra, Ganga, the central Indian River systems, the East and West flowing river systems of South India. River system constitutes of rivers, rivulets, tributaries, distributaries, gorges and nullahs. The Indus, Brahmaputra and Ganga together account for nearly 80 per cent of the total hydropower potential. In the case of Indus the utilization is, however, governed by the Indus Water Treaty with Pakistan. In India out of the total power generation installed capacity of 1,76,990 MW (June, 2011), hydro power contributes about 21.5 per cent *i.e.* 38,106 MW. A capacity addition of 78,700 MW is envisaged from different conventional sources during 2007-2012 (the 11th Plan), which includes 15,627 MW from large hydro projects. In addition to this, a capacity addition of 1400 MW was envisaged from small hydro up to 25 MW station capacities

In addition, pumped storage sites with an aggregate capacity to the tune of 94,000 MW have also been identified, but only about 5,000 MW have so far been developed. The assessment of small hydro (up to 25 MW) potential has indicated nearly 10,000 MW distributed over 4,000 sites. It is estimated there is still an unidentified small hydro potential of almost 5,000 MW. This is based on the reassessment (the first assessment was carried out during 1953-1959) of hydropower resources carried out in 1980s taking into account new information available on topographical features and hydrology of the river systems, technological advances and experience gained in civil works of hydropower projects, and the latest trends in relative economics of power generation from different sources. The assessment of energy is based on availability of water corresponding to a 90 per cent dependable year.

Major Hydro Power Plants in India

Hydro Power Plants in India

The country has developed numerous hydro power plants of various capacities. It is easier to build a hydropower plant where there is a natural waterfall. Dams are built on rivers where the terrain produce artificial lake or reservoir above the dam. Some of the main plants have been exhibited in the map to have their bird' eye view.

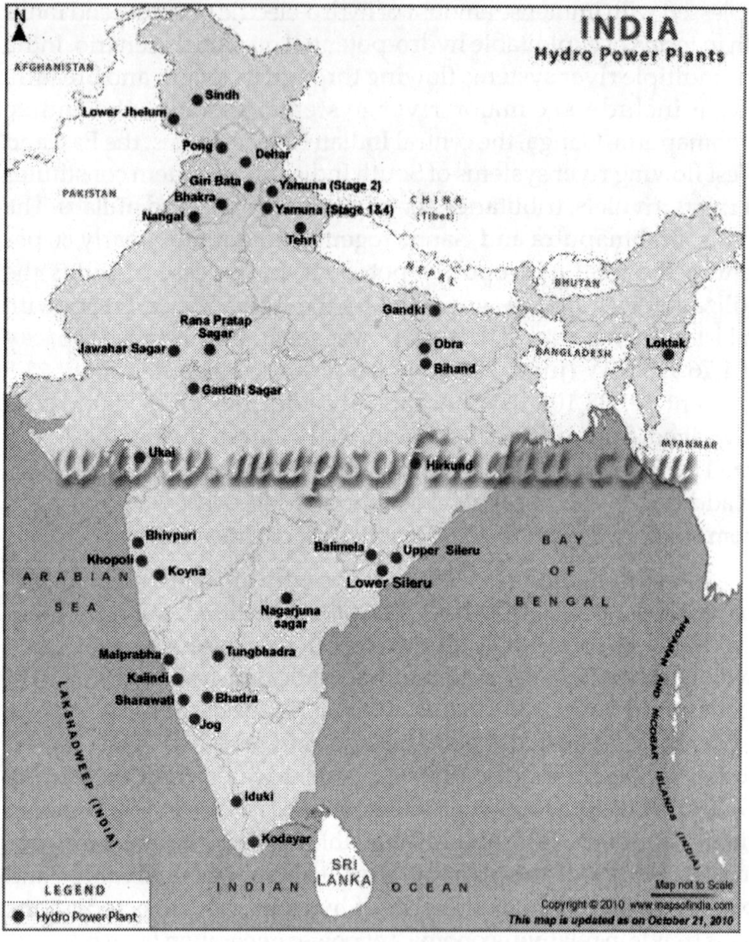

Map 21.1

Most of the plants located in the Northern India have been constructed on the rivers/rivulets originating from the Himalayas. These being snow and rain fed generate hydropower round the year and cater to domestic, commercial and industrial needs

Region-wise Potential

Regionally, the entire country is divided into five regions to have better understanding of the potential of hydropower. The information compressed in Table 21.1 portrays inter-regional state of assessed, developed and balance potential of hydropower. North-Eastern region has the highest assessed potential. But less than five per cent of it has been harnessed. Western and Southern India has tapped the higher proportion of hydropower. Still, almost three-fourth of the total potential is yet to be harnessed.

Table 21.1: Region-wise Potential of Hydro-Power*

Region	*Potential Assessed (MW)*	*Potential Developed (MW)*	*Potential under Development (MW)*	*Balance Potential (MW)*	*Balance Potential (per cent)*
Northern	30,155	5,150	2,905	22,100	73.28
Western	5,679	2,270	1,164	22,450	39.53
Southern	10,763	5,924	153	4,686	43.54
Eastern	5,590	1,364	201	4,025	72.00
North-Eastern	31,857	517	914	30,424	95.50
Total	84,044	15,225	5,339	69,480	75.53

*At 60 per cent Load Factor.

Source: Indian National Hydro Power Association.

Basin-wise Potential

Basin-wise Potential of Hydro-Power has been shown in Tables 21.2 and 21.3. The economically exploitable potential from these river systems through various schemes has been assessed at 84,044 MW at 60 per cent load factor corresponding to an installed capacity of around 148,700 MW. The country ranks fifth in terms of exploitable hydro-potential on global scenario. Brahmaputra Basin has the highest assessed and balance hydro potential but per cent could be harnessed. Eastern, Western and Central river basins have comparatively much less assessed potential but have exploited comparatively much higher proportion of it. Considering the total potential already developed and being developed, the Indus basin leads all other basins.

Table 21.3 depicts information of basins-wise hydro-electric potential at 100 per cent load factor in the country. In addition, 56 number of pumped storage projects have also been identified with probable installed capacity of 94 000 MW. In addition to this, hydro-potential from small, mini and micro schemes has been estimated as 6 782 MW from 1,512 sites. Thus, in totality India is endowed with hydro-potential of about 2,50 000 MW.

Table 21.2: Basin-wise Potential of Hydro-Power*

Basin	*Potential Assessed (MW)*	*Potential Developed (MW)*	*Potential under Development (MW)*	*Balance Potential (MW)*	*Balance Potential (per cent)*
Indus Basin	19,988	3,731	1,156	14,701	73.
Ganga Basin	10,715	1,901	1,367	7,447	69.
Central Indian Rivers	2,740	1,060	1,147	533	19.4
West Flowing Rivers	6,149	3,704	41	2,404	39.09
East Flowing Rivers	9,532	4,168	144	5,220	54.76
Brahmaputra Basin	34,920	661	1,085	33,175	95.00
Total	84,044	15,225	5,339	63,480	75.53

* At 60 per cent Load Factor.

Source: Indian National Hydro Power Association.

Table 21.3: Basin/River-wise Hydro-electric Potential

Sl.No.	*Basin/Rivers*	*Installed Capacity (MW)*
1.	Indus Basin	33,832
2.	Ganga Basin	20,711
3.	Central Indian River system	4,152
4.	Western Flowing Rivers of Southern India	9,430
5.	Eastern Flowing Rivers of Southern India	14,511
6.	Brahmaputra Basin	66,06
	Total	1,48,701

State-wise Installed Capacity

The data showing state and sector-wise installed capacity has been compressed in Table 21.4. Public and private sectors have embarked in generating hydropower in various states. About 19000

MW of electricity is being generated by central and state sectors together and nearly 12000MW by the private sector. Thus if central, state and private sectors are conserved as separate entities, the private sector has larger share in generating hydropower. Arunachal Pradesh has been producing highest hydropower (9579MW) followed by Uttarakhand (6858MW), Jammu and Kashmir (3923MW), Himachal Pradesh (2457MW) and Sikkim (2455MW) respectively. About two third of the total installed capacity by private sector in the country is being produced in Arunachal Pradesh alone.

Growth in Hydropower

There has been gradual increase in the installed capacity during various five year plans. The hydropower generation increased from

Table 21.4: State-wise and Sector-wise Installed Capacity

Sl.No.	*States*	*Central Sector*		*State Sector*		*Private Sector*		*Total*	
		No.	*MW*	*No.*	*MW*	*No.*	*MW*	*No.*	*MW*
1.	Himachal Pradesh	2	816	7	892	6	749	15	2457
2.	Jammu & Kashmir	4	2450	4	1473	0	0	8	3923
3.	Uttarakhand	12	4374	7	1655	5	829	24	6858
4.	Punjab	0	0	1	168	1	75	2	243
5.	Madhya Pradesh	3	166	0	0	0	0	3	166
6.	Andhra Pradesh	0	0	3	1560	0	0	3	1560
7.	Kerala	0	0	6	373	0	0	6	373
8.	Karnataka	0	0	2	400	0	0	2	400
9.	West Bengal	1	120	2	66	0	0	3	186
10.	Sikkim	1	520	0	0	10	1935	11	2455
11.	Arunachal Pradesh	3	1610	0	0	23	7969*	26	9579
12.	Assam	0	0	1	150	0	0	1	150
13.	Manipur	2	1566	0	0	0	0	2	1566
14.	Tamil Nadu	0	0	1	500	0	0	1	500
15.	Meghalaya	0	0	1	54	1	450	2	504
	Total	28	11622	35	7291	46	4038	109	30920

Includes only 600 MW out of 2400 MW Siang Lower H.E. Project for capacity addition during 12th plan.

Contd...

Table 21.4–Contd...

Plan-wise Growth and Share of Hydro-Power

Plant Period	*Installed Capacity at the End of Plant (MW)*		
	Hydro	*Total (Including Renewable)*	*Hydro as Per cent of Total*
1st Plan (1951-56)	1061	2886	36.78
2nd Plan (1956-61)	1917	4653	41.19
3rd Plan (1961-66)	4124	9027	45.68
Three Annual Plants (1966-69)	5907	12957	45.58
4th Plan (1969-74)	6966	16664	41.8
5th Plant (1974-79)	108.33	26680	40.6
Annual Plan	11384	28448	40.01
6th Plan (1980-85)	14460	42585	3.96
7th Plan (1985-90)	18307	636636	28.77
Two Annual Plans (1990-92)	19194	69065	27.79
8th Plan (1992-97)	21658	85795	25.24
9th Plan (1997-02)	26269	105046	25

1061 MW at the end of first plan to 26269 MW at the end of ninth plan. However, share of hydro-electricity decreased from nearly 46 per cent in third plan to 25 per cent by the end of ninth plan. There has been descendance in the share of hydropower to the total renewable power.

Inter-regional Plans for Power Generation

India is pursuing a centralized system for power planning. These plans would be continuously updated to take care of the revisions in load projections and generation scenarios. Further detailed planning by the Central and State transmission utilities has to conform to this plan. A draft National Electricity Policy (NEP) was notified in 2005 which was based on an all-India generation capacity addition of around 40,000 MW during 2002-2007 and the demand projections as shown in Table 21.5.

Table 21.5: Region-wise Forecasts of Power

Region	*Energy Requirement (MW)*		*Peak Load (MW)*	
	2011-2012	*2016-2017*	*2011-2012*	*2016-2017*
Northern	308,528	429,480	49,674	69,178
Western	299,07	395,859	46,82	61,966
Southern	262,718	354,599	42,061	6,883
Eastern	90,396	117,248	15,664	20,416
North-Eastern	14,061	20,756	2,789	4,134
All India	975,222	1,318,644	157,107	212,72

The plan covered two scenarios of GDP growth rate; namely, 6.5 per cent and 7.4 per cent for working out generation capacity additions. Besides, the likely impact of various factors such as limited indigenous coal availability–367 metric tons (MT) as compared to 419 MT in the base case–energy conservation (about 6.9 per cent), peak reduction (5 per cent), inter-regional diversity in demand (3.5 per cent), accelerated rural electrification, and household modernization, spinning reserve requirement (5 per cent), etc. and low hydro scenario (benefits of 5,000 MW of hydropower slipping from the 11th Plan to the 12th Plan) have also been studied. Besides, the likely impact of various factors such as limited indigenous coal availability–367 metric tons (MT) as compared to 419 MT in the base case–energy conservation (about 6.9 per cent), peak reduction (5 per cent), inter-regional diversity in demand (3.5 per cent), accelerated rural electrification, and household modernization, spinning reserve requirement (5 per cent), etc. and low hydro scenario (benefits of 5,000 MW of hydropower slipping from the 11th Plan to the 12th Plan) have also been studied.

Bottlenecks in Hydro-development

Many hurdles come in the way of development of hydropower projects. Longer gestation period and capital intensive nature of the projects: Preparation of detailed projects reports (DPRs) for hydropower projects takes relatively longer period than for thermal projects because reliable hydrological, geological, seismological and environmental studies have to be carried out for a longer period. Thus hydropower projects generally entail a long gestation period.

In addition to this, these projects are comparatively capital intensive. In the context of resource shortages and continuing power shortages, thermal projects (coal, liquid fuel and gas), which need a relatively short gestation period, have been getting priority in fund allotments.

Besides, hydro development is plagued by several constraints which have affected hydro development. Some of such constraints are as under:

1. **Technical**: Investigation, inadequacies in tunnelling methods;
2. **Financial:** Deficiencies in providing long term finance;
3. **Tariff:** Tariff and other related issues;
4. **Managerial:** Poor contract management;
5. **Geological**: Especially in the Himalayan region where underground tunnelling is required;
6. **Topographical**: Inaccessibility of the area;
7. **Miscellaneous**: Problems due to delay in land acquisition, and resettlement of project affected families, law and order problem etc.

Suggestions to Accelerate Hydropower Development

The Government of India has set the following objectives for accelerating the pace of hydropower development:

1. Ensuring targeted capacity addition during various plans.
2. Exploitation of vast hydro-electric potential at a faster pace.
3. Promoting small and mini hydel projects.
4. Strengthening the role of PSUs/SEBs for taking up new hydel projects.
5. Increasing private investment.

The public sector organisations can play a greater role in the development of new schemes, this alone would not be adequate to develop the vast hydro potential since it will require huge investments which are difficult to be supported from the budget/ plan assistance in view of competing demands from the various sectors. A greater private investment through joint ventures would be encouraged in the coming years and required atmosphere, incentives and reliefs would be provided to stimulate and maintain a trend in this direction.

Policy for Hydropower Development

Life span of hydro-electric projects extends beyond 50 years. The projects open avenues for development of remote and backward areas help in conserving scarce fossil fuels. Our country has enormous economically exploitable and viable hydro potential assessed to be about 84,000 MW at 60 per cent load factor (148,700 MW installed capacity). In addition, 6781.81 MW in terms of installed capacity from small, mini and micro hydel schemes have been assessed. Also, 56 sites for pumped storage schemes with an aggregate installed capacity of 94,000 MW have been identified. However, about one fifth of the hydro-electric potential has been harnessed so far and 7 per cent is under various stages of development. Thus, nearly three-fourth of the potential remains without any plan for exploitation. Despite hydro-electric projects being recognised as the most economic and preferred source of electricity, share of hydro power has been declining steadily since 1963. The ideal hydro thermal mix should be in the ratio of 40:60. Because of an imbalance in the hydel thermal mix especially in the Eastern and Western regions, many thermal power stations are required to back down during off peak hours. The capacity of the thermal plants cannot be fully utilised resulting in a loss of about 4 to 5 per cent in the plant load factor. To achieve the objective for faster development of hydro potential, the Government proposes to take the following steps and measures:

Funding

Having regard to the large capacity addition envisaged in the state sector (5810 MW) it is necessary to (a) provide a mechanism for funding hydro projects by earmarking funds in the plan allocation of the State Governments by the Planning Commission; and (b) organising supplementary funding of hydel projects where more than 50 per cent of the expenditure has already been incurred.

Power Development Fund

Project Reports are offered to private developers, their response could be more favourable. If pre-construction activities and enabling works could be completed and these sites offered to IPPs

(Independent Power Producers) the chances of IPPs opting to invest in these projects would further improve. Further this would reduce the gestation period which would make investment in hydro projects more attractive. It is, therefore, proposed to levy a Power Development Cess at 10 paise per kwh of electricity consumed in the country. The levy of cess was recommended by the Sub Committee of the NDC Committee on Power which gave its report in January, 1994. The cess would be levied on the electricity billed by (State Electricity Boards) SEBs/Electricity Departments/Bulk licensees/Distribution licensees. The State/UT Governments would be responsible for the collection of the cess. The amount would thereafter be credited to a "National Power Development Fund".

Basin-wise Development of Hydro Potential

The assessment of hydro potential in 84 identified conventional hydro projects and 56 pumped storage projects is on the basis of desk studies using toposheets and discharge data. Further, detailed studies to firm up the parameters of the projects as identified by Central Electricity Authority (CEA) would be taken up on the basis of development of hydro potential in a basin as a whole for maximising benefits and prioritising execution of projects. These studies will be done in close co-ordination with CWC and Planning Commission and in harmony with development for other uses of water like irrigation, drinking water etc. While CEA would carry out these studies, (Central Public Sector Undertakings) CPSUs/other Central Government Organisations and State. Authorities would do the investigations and prepare the detailed project reports, by adopting an integrated approach towards planning and development of the various projects, evacuation arrangement and environmental impact assessment. This would enable an optimal harnessing of hydro potential in each river basin.

Advance Action for Capacity Addition

Government has taken immediate steps to tie up funding, execution agencies and convey investment decision for schemes already accorded techno-economic clearance of CEA. As far as Central Sector is concerned, NHPC Chamera Stage II has taken up (300 MW), Parbati Stage-II (800 MW) in HP; Teesta Stage V (510 MW) in Sikkim, Loktak Downstream (90 MW) in Manipur and NEEPCO has taken up Tuivai (210 MW) in Mizoram, Lower Kopili

(150 MW) in Assam, Kameng (600 MW) and Ranganadi Stage II (160 MW) in Arunachal Pradesh. In addition Tehri Hydro Development Corporation Ltd (THDC) would take action to start activities on Tehri Stage II (1000 MW) and Koteshwar (400 MW) in Uttarakhand. Similarly Satluj Jal Nigam Ltd (SJNL) has taken up Rampur Project (410 MW) in HP.

Survey and Investigations

As a long term strategy efforts will be made to ensure that DPRs which are under various stages of processing for accord of TEC by CEA are finalised and cleared so that a start could be made on these projects in the next one or two years. Survey and investigation of the potential hydro sites on an advanced scientific basis would be essential requirement for the future. The progress on this front has been dismal given the funds constraint and outdated technology. The funding agencies like World Bank and (Asian Development Bank) ADB have shown their interest toward funding the survey and investigation activities for hydro-electric projects. Concerted efforts would be made towards availing the funds quickly.

Inter-State Projects

A substantial hydel power potential has remained locked up and many mega hydel projects could not be taken up for implementation, even though these projects are well recognised as attractive and viable, because of unresolved Inter-State issues. Govt. of India recognises the need for evolving an approach to ensure that the available hydro-electric potential is fully utilised without prejudice to the rights of the riparian States as determined by the Awards of the Tribunals/Agreements arrived at among the party States for a given river basin with regard to water sharing. The selection and design of project would be based on integrated basin wise studies, so as to arrive at an optimal decision and care will be taken that such projects do not in any way prejudice the claims of basin states or affect benefits from the existing projects.

Renovation, Modernisation and Updating

Renovation, modernisation and uprating of old hydro power plants is being accorded priority as it is a faster and cheaper way of capacity addition than installing new capacity.

The Government of India has set the target for India's optimum power system mix at 40 percent from hydropower and 60 percent from other sources. It is high time for India to give primacy to Green Power and Hydro Power.

References

Worldwatch Institute (January 2012). "Use and Capac it y of G l oba l Hydropower Increases". ht t p: / / www.wor l dwa t ch.org / node / 9527.

Braz il Sends Forces to Jirau Dam Af t er R i o t s". Wall Street Journal. 18 March 2011. ht t p: // onl i ne.wsj.com / ar ti c l e / SB10001424052748704608504576208640384691826.htm l. Retrieved 2 April 2011.

Batta, R.N. (2000). *Tourism and the Environment: A Quest for Sustainability*. Indus Publishing. ISBN 8173871108. http:// books.google.com/?id=lDJrdjwM7qYC and dq=himachal. Retrieved 2012-03-11.

Singh, Mian Goverdhan (2000). C.L. Gupta and Kulbhushan Chandel. ed. *Himachal Pradesh: History, Culture and Economy* (extensively revised ed.). Shimla.: Minerva Publishers and Distributors.

Singh, Sarina (2005). *India*. Lonely Planet. ISBN 1740596943. http:/ /books.google.com/?id=Fk8FQa2ZSFQC and dq=himachal+ history. Retrieved 2012-02-11.

Tiwari, A.K. (2000). *Infrastructure and Economic Development in Himachal Pradesh*. Indus Publishing. ISBN 8173871051. http:// books.google.com/?id=aaXQwI5j9-MC and dq=himachal. Retrieved 2011-09-31.

Verma, V. (1995). *The Emergence of Himachal Pradesh: A Survey of Constitutional Developments*. Himachal Pradesh (India): Indus Publishing. ISBN 8173870357. http://books.google.com/ ?id=QpWloqN5LTAC and dq=himachal+history. Retrieved 2011-06-15.

United Nations Development Programme (UNDP) (2002). *Himachal Pradesh Human Development Report 2002*. Government of Himachal Pradesh. ISBN 019521915. http:// www.undp.org.in/index.php?option=com_content and task=view and id=128 and Itemid=216. Retrieved 2011-08-21.

Patrick James, H Chansen (1998). "Teaching Case S t udies in Reservoir S ilt a ti on and Ca t chment Erosion". Great Britain: TEMPUS Publications. pp. 265–275. ht t p: / / www.i j ee.di t.i e / ar ti c l es/Vol14-4/ ij ee1012.pdf.

Rediscovered" Wood and The Trit on Sawfi sh". Inhabitat. 2006-11-16. ht t p: //i nhabi t a t.com / 2006/12 / 01/rediscovered-wood- t he- t r it on-sawf i sh/#more-1973.

Briefing of World Commission on Dams". Internationalrivers.org. 2008-02-29. ht t p: //i nternationalri vers.org /en/way-forward/world-commission-dams/world-commission-dams-framework-brief-introduction.

Index

Z